PRAISE FOR *WHAT C*

Former Harvard neurosurgeon Eben Alexander, MD, author of the *New York Times* bestseller *Proof of Heaven: A Neurosurgeon's Journey into the Afterlife* – "A clearly written book showing how quantum physics supports an afterlife...The science of quantum physics, born in the heyday of materialist science, has spent the last century trying to reveal to the scientific community the fundamental, unified nature of consciousness in the universe. The real breakthroughs have involved serious investigation of many aspects of human experience (such as near-death experiences, telepathy, distance healing, remote viewing, etc.) that, when combined with the deep lessons of quantum physics as they are in journalist Michael Schmicker's book, awaken us to the profound implications of this demonstration of the primacy of mind. The afterlife is not only allowed, but is a necessary ingredient for making sense of this expanded scientific world view."

Physicist Fred Alan Wolf, author and National Book Award winner in Science for *Taking the Quantum Leap: The New Physics for Nonscientists* – "*What Comes Next?* is very well-written, and covers an amazing deal of ground concerning the "big question" we all ask ourselves. Quantum physics has pulled from under us the materialist rug that we physicists have long stood on. It has made most of us realize that any materialist answer to 'what comes next' must be false…"

Dr. Jeffrey Long, MD, author of the *New York Times* **bestseller,** *Evidence of the Afterlife: The Science of Near-Death Experiences,* **and founder of the Near-Death Experience Research Foundation** – "This captivating book brilliantly explains how quantum physics supports the findings of Near-Death Experience research. It blends scientific inquiry and the author's personal journey to answer one of humanity's most enduring questions: Does consciousness survive death? It proposes an intriguing possibility backed by hard scientific evidence – yes, an afterlife is both scientifically possible and logical under the rules of quantum physics. My enthusiastic recommendation."

Psychologist Jeffrey Mishlove, PhD, host of the national public television series *Thinking Allowed* **and the YouTube channel** *New Thinking Allowed,* **and author of** *The Roots of Consciousness.* – "The scientific evidence for life after death is substantial, as Michael Schmicker points out. For the most part, however, this enormous body of evidence has been marginalized by the scientific and educational establishments. Mostly, it is thought that the mountains of data accumulated in the fields of psychical research, consciousness research, and parapsychology must be flawed – because it is assumed that survival after death is inconsistent with physics. This book is important because it shows that such an assumption is not warranted."

Physicist Nick Herbert, author of *Quantum Reality* – "Is death a dreamless sleep or something grander? Addressing this question not from a religious or mystical perspective but from the imaginative efforts of scientists probing this terminal event as closely as possible without themselves falling in, Michael Schmicker assembles evidence from many quarters, including hints from quantum theory, that the end of consciousness may not be as simple as some of us imagine. A stimulating, accessible, and optimistic book."

Dean Radin, PhD, Chief Scientist, Institute of Noetic Sciences, and author of *Entangled Minds* – "*What Comes Next*, by investigative journalist Michael Schmicker, explores the relationship between quantum mechanics and consciousness, leading to a surprising affirmation of the afterlife. This well-written book bridges the gaps between science and spirituality, compelling readers to rethink the role of consciousness in the physical world and offering a refreshingly optimistic answer to the perennial questions about life after death."

Keith Parsons, British documentary film producer of *This Life, Next Life* – "...a brilliant piece of work involving enormous research, summarized very effectively."

What Comes Next?

An Investigative Reporter
Uncovers Quantum Physics'
Hidden Afterlife Hypothesis

Michael Schmicker

PALLADINO BOOKS

What Comes Next? An Investigative Reporter Uncovers Quantum Physics' Hidden Afterlife Hypothesis. Copyright © 2024 by Michael Schmicker. All rights reserved. Printed in the United States of America. No part of this book may be used or reproduced in any manner whatsoever without written permission except in the case of brief quotations embodied in critical articles and reviews. An imprint of Palladino Books. First Palladino Books paperback edition published in 2024.

ISBN: 9798325288746
Visit www.MichaelSchmicker.com
Cover: Andy Carpenter/ACD Book Cover Design
E-Book available online from Amazon.com
Audiobook available from Amazon.com

CONTENTS

Introduction	11
Part I: The Investigation	15
The Pam Reynolds Lowrey NDE	23
The Afterlife State	26
The Hard Problem in Neuroscience	31
The Quantum Physics Revolution	35
The Afterlife: 19th-Century Newtonian Science	37
The Afterlife: 21st-Century Quantum Physics	40
An Afterlife is Scientifically Possible and Logical	44
Does our Consciousness Create the Universe?	47
Theater of the Mind	53
Five Surprising Scientific Conclusions	55
Part II: The Scientific Evidence	61
NDE Research Says Consciousness Survives Death	65
Neuroscience Can't Solve the "Hard Problem"	87
Reality Isn't Made of Newtonian Matter	113
Newtonian Scientific Materialism is False	121

Does Consciousness Create the Universe?	139
Why Science Ignored Quantum Physics	151
Quantum Physics and the "Paranormal"	161
Digging Deeper Into Quantum Physics	167
Nick, Gene, Me – and You	187
About the Author	193
Acknowledgements	195
Photo/Art Credits	197

DEDICATION

This book is dedicated to physicists Nick Herbert, Fred Alan Wolf, and their colleagues in the celebrated Fundamental Fysiks Group, which met at the Lawrence Berkeley National Laboratory in California in the 1970s. Their freewheeling, maverick thinking forced mainstream physicists to pay attention to the strange but exciting implications of quantum theory for our understanding of reality.

INTRODUCTION

I'm getting up there in age. I still jog two miles a day, work through midnight to meet a deadline, and weigh just nine pounds more than I did back in college. But the Social Security actuarial tables don't lie. My life expectancy is clearly shrinking fast.

I find myself unquestionably old – but surprisingly cheerful. I'm still entranced by the sheer beauty and delightful complexity of this earthly life in all its forms, from maple trees and stars to the infinite variety of humanity's creativity, thoughts, dreams, beliefs, and rich emotions. In my humble opinion, we humans are really something wonderful. Sometimes I catch myself laughing out loud.

When you get to be my age, friends start asking you your opinion about the meaning of life. "Where did we come from? Where are we going?" Most of all, they want to know what I think happens when we die. What comes next? Is there an afterlife?

They know I'm an investigative reporter; I've written books about scientific anomalies, including near-death experiences, deathbed visions, and reincarnation claims. Surely, I've given it some thought? I have.

I'm happy to report that, late in life, I've found – at least for myself – a hypothesis that, to my surprise and delight, is both hard-grounded in science and astonishingly optimistic. In a nutshell, quantum physics suggests that our consciousness survives death, and the physical universe is filled with purpose and meaning because we literally help form this universe.

I didn't come up with this hypothesis, nor is it some New Age woo-woo. Multiple Nobel Prize–winning scientists working in quantum physics did. I simply stumbled across it late in life. I realize most people are too busy or too disinterested to follow scientific research aimed at understanding the nature of reality. I don't blame them. We've all got a million things to do. But when you're ready, the scientific evidence is there, waiting for you.

Nineteenth-century Newtonian science closed the door on an afterlife.

Twenty-first century quantum physics reopens it.

This scientific hypothesis has made a huge difference in how I see my value and my place in the universe.

Maybe it will for you too.

One last note before we start: This book is divided into two very different sections.

Part I is simply the personal story of my investigation, told with a minimum of science.

Part II contains the heavy reading. It's the hard scientific evidence supporting the afterlife argument I make in this book (no belief required. It's simply current science):

Near-Death Experience research says our consciousness survives death. Quantum physics explains why this is scientifically possible and logical.

Part II is aimed at skeptics (like me) who require evidence: blue-chip scholarly experts; credible, vetted citations; and step-by-step deductive reasoning. They'll find them here. But if you're a person who trusts your intuition over academic arguments when it comes to deciding what is true or not, you may decide to skim or skip Part II.

PART I: THE INVESTIGATION

For most of my life, I had to choose between two hypotheses:

The Religious Hypothesis: God created us and the physical world. There is an afterlife.

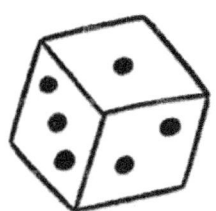

The Newtonian Science Hypothesis: Blind chance created us and the physical world. There is no afterlife.

I grew up with the Religious Hypothesis, given to me at birth, and based on faith. I accepted it with minimal questioning, right up to adulthood. My parents' Roman Catholicism was filled with rules, but it wasn't some frightening, sour, fundamentalist version of Christianity. It provided multiple benefits – a meaning to life (death is not the end), a chance to make mistakes and easily recover (via confession), and an optimistic outlook on daily life (a loving God is in control).

But the longer you live, the more exposure you get to other people's beliefs and ideas. In college, I took a comparative religion course. I quickly realized there were a hundred different competing religions out there to choose from. How do you choose among them – or *for* any of them? At that point, I made the first important adult decision of my life – I would decide what was true and real based only on the evidence available for it. "Show me the evidence" became my guiding rule for judging any claim, made by anybody, about anything.

I looked hard, but I couldn't find any good, scientific evidence for the heaven promised by my childhood religion – Newtonian science declared an afterlife was a decisively disproven idea, unworthy of serious attention or investigation. I found no evidence for Christianity's belief in the existence of Adam and Eve, or Original Sin (the reason Christianity gives for us needing salvation), or a virgin birth, a flaming hell, eternal damnation, a creature called Satan, an archangel Michael, an infallible Pope or infallible Bible. Other religions didn't fare any better. I found no scientific evidence for the Islamic belief that Mohammad was God's only prophet, or Judaism's belief that they were God's Chosen People, or Hinduism's belief that the god Brahma created the universe.

So I abandoned religion and adopted the Newtonian science hypothesis. The Newtonian science view of reality had none of the comforts of my childhood religion. But at least it offered some evidence for its claims. Besides, focusing only on what could be physically measured, weighed, and manipulated worked magnificently in daily life. We created amazing technologies that made material life less dangerous and more pleasant in so many ways, for so many people – a safer, healthier, better-fed, more educated world. I don't know about you, but I appreciate those improvements. If there was no afterlife, at least we lived better in the one life we had.

But it's also depressing. Nineteenth-century Newtonian classical physics is grounded on the philosophy of Materialism. The only thing real is the physical world of measurable, tangible objects, knowable through our five senses. It's a universe created by blind chance, following impersonal, mechanical laws. There's no grand purpose to our universe. There's nothing special about us. We're simply a temporary collection of chemicals and atoms. Cut humans open, put them under a microscope, and you won't find such a thing as a soul. Consciousness is created by the brain, by the firing of neurons. When the brain dies and the firing stops, our consciousness – our self-awareness, thoughts, memories, hopes, dreams, feelings – are permanently annihilated. It's as if we never existed in the first place.

Not that I gave up looking for contrary scientific evidence. I'm a voracious reader with an open mind. I studied philosophy in college, which taught me to question everything we think we know about reality.

I loved it. You may have heard this one – the Chinese poet and philosopher Chuang-Tzu fell asleep in a boat on a lake one summer day. He dreamed he was a butterfly. When he awoke, he asked himself the question, "Am I a man who dreamed he was a butterfly? Or am I a butterfly dreaming I am a man?" Think about it. How do you *scientifically* prove you're not a butterfly?

After college, I became an investigative journalist, which reinforced my tendency to be skeptical about everybody and everything.

When I was 25 years old, the life-after-death question suddenly became urgent and personal. I got a notice I was being drafted into the army to fight in Vietnam.

College friends of mine were already dying there. So, was this the meaning and purpose of my life? By pure chance (because I was born male and turned draft age at just the wrong time), to be  sent to Asia in my twenties, to possibly be killed in an unnecessary, senseless war? Per the Newtonian science hypothesis, yes.

At the last minute, I received a surprise two-year deferment to teach English in Thailand for the Peace Corps. Thai people believe in the  paranormal. It's a land of ghosts and spirits, gold amulets worn around your neck to protect you from enemies, fortune tellers and astrologers to predict your future. All my fellow teachers at the Buddhist monastery school where I taught English accepted the reality of a spirit world. I tried to argue with them. "Show me the evidence, and I'll consider it," I said. My Thai friends just smiled and shrugged. They had experienced these things. Maybe someday I would too. Meanwhile, they said to me, it was a beautiful, sunny day, the *khao pad* (Thai fried rice) was delicious, and we only had twenty minutes for lunch.

A year after I returned to the States, I overheard at a family party that one of my Swiss relatives, a Catholic priest named Father Eugene Weibel, had had multiple ESP experiences. I was fascinated. He had no motive to lie, nor was poor memory involved – he logged the experiences in his daily diary.

The Newtonian science hypothesis said he was deluded, crazy, or lying, because telepathy was impossible. Until then, I had never bothered to examine the scientific evidence for or against ESP. Now I had a reason to.

I had a master's degree; I knew where to find scientific papers. I was also a journalist. I knew how to track down leads, to interview experts. In my spare time, I did an intensive search for the best scientific evidence available for ESP. I was surprised by what I uncovered. I expected to find some evidence, but I never expected to find evidence of such quality. It included a decade of pioneering telepathy experiments at Duke University; the existence of a reference database containing 14,000 anecdotal ESP reports, providing scientists with information about patterns of human ESP experiences, as well as suggestions for testable hypotheses; and successful ESP tests conducted under extremely tight guidelines created by a panel of die-hard skeptics (including professional magicians, who signed off on the test protocols, and the computer-controlled selection of test targets, to eliminate any human bias). The conclusion? ESP was quirky but real, and not explainable by blind chance.

Intrigued and curious, I expanded my research effort to find the best scientific evidence for other so-called "paranormal" phenomena reported by humans for centuries, including psychokinesis (then being researched at Princeton University's School of Engineering); mental healing (Harvard University); near-death experiences, deathbed visions, and reincarnation reports (University of Virginia School of Medicine); and after-death communications via mediums (University of Arizona). In 2000, I published my findings in a book called *Best Evidence*.

During my book research, I kept coming across skeptics who refused even to look at the evidence. Their primary argument seemed to be "these phenomena can't exist; therefore, they don't." They reminded me of the church cardinals refusing to look through Galileo's telescope, afraid they'd see evidence that would contradict their beliefs. Dogmatic scientists are no different from dogmatic clerics. What most

upset many of them was the scientific evidence suggesting the possibility of life after death. A university biology professor emailed me to praise *Best Evidence* for its "impressive scholarship" but wouldn't recommend it to his students. "It lets religion slip in the back door of science."

Ironically, I was getting hammered on the internet by religious fundamentalists. Why? Because the evidence from reincarnation cases and near-death experiences doesn't support their afterlife belief in a wrathful, judging God or eternal damnation. Even more upsetting? Though some NDE experiencers did report meeting Jesus, others didn't. They met the Indian god Yama; or the Buddha; or a nameless Being of Light that enveloped them in love; or no god at all, just deceased relatives.

Conversely, nihilists hated the evidence. You expect nothingness. Instead, you're still here. One of the funniest NDE reports I read involved a woman who didn't believe in an afterlife. She suffered a terrible car accident and found herself floating above her lifeless body. Looking down at the dress she was wearing, she was upset at the thought she might have to spend eternity in a dress she hated.

The near-death-experience (NDE) research intrigued me the most. Everyone wants to know what happens when we die, me included. According to 19th-century Newtonian science, when the brain and heart stop working, our consciousness is extinguished. Yet physicians and medical researchers in the US and Europe were reporting cases of people continuing to be conscious even in the state of verified clinical death. Skeptics have offered some twenty-plus different arguments to try and explain NDEs away, but NDE researchers have strongly challenged them.

I included one of the strongest (and most famous) NDE cases in *Best Evidence*. American singer-songwriter Pam Reynolds Lowrey underwent a risky brain operation at Barrow Neurological Institute in Phoenix, Arizona, in 1991. Neurosurgeon Robert

Spetzler put her in full, hypothermic cardiac arrest. They dropped her core body temperature to 60 degrees, completely drained all the blood from her brain, and left her with no measurable heartbeat or brain waves. Flatlined EKG and EEG monitors verified her state of clinical death. During this state of death, the woman found herself fully conscious, out of her body, floating above the operating table where she reported looking down and seeing the surgeon cutting her skull with an odd-looking Rex bone saw. She also heard nurses talking about how small her veins and arteries were. After watching this for a while, she continued through a now-classic NDE experience – moving through a dark tunnel, seeing a light, meeting and communicating with her dead grandmother and uncle, and finally returning to her body on the operating table. Afterwards, Reynolds at first assumed she had been hallucinating. But a year later, she mentioned the details to her neurosurgeon. Spetzler said her account matched his memory. Spetzler did not check out all the details. But Atlanta cardiologist Michael Sabom, doing research on NDEs, did. With Reynolds' permission, Spetzler sent Sabom the records from the surgery. The records confirmed Reynolds had no electrical activity in the brain (no neurons firing) during the time she said she was conscious and floating above the operating table observing the bone saw operation. Likewise, at that time, she had no sight or hearing (her eyes were taped shut, and her ears were plugged). Yet her description of what happened during the operation was both accurate and extremely detailed (it couldn't be explained by lucky guessing).

In a May 22, 2009, interview on National Public Radio (NPR), Spetzler confirmed that Reynolds' account matched his memory, while adding, "From a scientific perspective, I have absolutely no explanation about how it could have happened."

If you've never heard of this case, join the crowd. It requires interest, time, and effort to follow the field of near-death-experiences research. Meanwhile, you've got to go to work, fix the car, find a Starbucks. But please know the scientific evidence is there, waiting for you, when you're ready to look.

Ironically, while researching near-death experiences for *Best Evidence*, I learned, to my great surprise, that my sister Rosemary had almost died from blood loss while giving birth to her daughter. During the crisis, she experienced a classic NDE. She never told anyone, she confessed, because she was afraid people would think she was crazy (it wasn't until 1975 that psychiatrist Raymond Moody came out with the first book on the subject).

What could life be like in a non-physical, afterlife state?

NDE reports are anecdotal, and they cover only a short time in the verified after-death state (hours at most), but they do offer us a starting point for trying to develop a conceptual model of the afterlife experience.

Psychiatrist and NDE researcher Bruce Greyson is professor emeritus of psychiatry and neurobehavioral sciences at the University of Virginia and a giant in the field of NDE research.

In 1978, he co-founded the International Association of Near Death Studies (IANDS) and for twenty-seven years edited the *Journal of Near-Death Studies*, the only scholarly journal dedicated to near-death research. He's also a Distinguished Life Fellow of the American Psychiatric Association (the highest honor bestowed by that organization).

In his book *After: A Doctor Explores What Near-Death Experiences Reveal About Life and Beyond*, he includes a chapter summarizing what NDE experiencers have reported regarding conditions they found in the non-physical state.

Here are some highlights:

- *Where is the afterlife?* It's not necessarily a physical "place," but rather a state of consciousness. Some 75 percent of experiencers say they went to an unfamiliar realm or dimension, without putting a label on it. Most said it couldn't be described in words. When pressed, they used a variety of metaphors ranging from religious terms like "heaven" or "hell," to nature terms like "a beautiful valley" or "a meadow," or to "outer space." But even when pressed, almost half insisted they still couldn't find *any* familiar label at all that fit it.

- *Is it pleasant or unpleasant?* Some 86 percent said it was primarily pleasant, 8 percent reported it was unpleasant, and 6 percent said neither.

- *Do you meet deceased family and friends?* In the afterlife state, some do, some don't. Sixty-six percent of NDE experiencers report meeting at least one other person. Two-thirds of those meetings involved meeting a deceased person. If you do meet someone who has passed on, they are overwhelmingly likely to be family. A 2001 study by Greyson's colleague, Emily Williams Kelly of the Department of Psychiatric Medicine at the University of Virginia, found that 95 percent of the deceased individuals encountered by NDE experiencers were relatives; only 5 percent were friends or acquaintances. Jeffrey Long, MD, created the Near-Death Experience Research Foundation (NDERF), which boasts over 1,300 NDE cases from around the globe – the largest NDE database in the world (see Part II, the "NDE research strongly suggests that consciousness survives death" section). Long's data shows that reunions are almost always joyful, not like the scary encounters seen in a ghost movie. Per

Long: "Although many deceased loved ones prior to death were elderly and sometimes disfigured by arthritis or other chronic illnesses, the deceased in the near-death experience are virtually always the picture of perfect health and may appear younger – even decades younger – than they did at the time of their death. Those who died as very young children may appear older. But even if the deceased appears to be a very different age than when they died, the NDE experiencer still recognizes them."

- *How do you communicate?* According to Long's research data, non-verbal telepathy is the type of communication that takes place during almost all near-death experiences in which communication is described.

- *Will you meet "God"?* You will likely meet a "divinity" of some sort, but it may not be the "God" you expect. Some 90 percent report encountering some kind of divine or godlike being. A minority, 33 percent, of experiencers identify the being they encounter with the God or leader of their particular religion – a bearded, blue-eyed Jesus; Buddha; Krishna;

Allah; the Chinese goddess Kwan Yin; the Celtic deity Cernunnos. But the majority (66 percent) don't meet a specific religious deity at all. Instead, they encounter an intelligent, non-specific being described as a Light, a Presence, or an all-loving deity for whom the term "God" is too puny. Even some avowed atheists report encountering a divine-like presence. Many experiencers come away from their NDEs concluding we are *all* divine. Greyson personally struggled with the results. He was raised in a scientific household without a strong sense of the divine. "I was uncomfortable with the overwhelming number of experiencers who described meeting some kind of god-like being."
But, he says, "science can't pick and choose which evidence is worth pursuing and which can be ignored. If we claim to be skeptics, we can't reject the observations that contradict our worldview, and accept those that agree with our views, without looking at the data. As Sigmund Freud warned us, 'If one regards oneself as a skeptic, it is a good plan to have occasional doubts about one's skepticism.'"

- *Most importantly, do you remain "you"?* Yes. While your senses may be altered, your personality and your memories remain. You continue to think, feel, and remember, just like you do in your physical life. You remain "you."

Beyond this limited scientific data, we largely enter the world of religious belief or philosophical speculation.

Does the afterlife last for eternity? What do we do in the afterlife? Do we play? Sleep? Learn? Move on to a higher state? Reincarnate? Voltaire once wrote, "It is no more surprising to be born twice than once." For a sample of the serious scientific evidence for reincarnation, try the book *Before* by Jim Tucker, MD, Professor of Psychiatry and Neurobehavioral Sciences at the University of Virginia. Do dogs go to the afterlife too? This one particularly interests me. If not, I've got a little dog, Cassie, whom I will miss greatly.

Meanwhile, after almost a century of trying, classical 19th-century Newtonian science still can't explain, much less scientifically prove, how or why the physical brain – a three-pound bag of fat, water, protein, carbohydrates, salts, and electrical impulses – can produce our subjective, inner, personal experiences.

We experience self-awareness (I exist); we think, plan, analyze, choose; we feel love and spiritual/mystical epiphanies; we feel empathy for others; we feel hope and dread, happiness and sadness, joy and despair.

Atoms don't.

If we're only made of atoms, why do we?

It's famously called the "hard problem" in neuroscience.

In July 2023, believers in the assumption that "mind is a product of the brain" suffered a spectacular public embarrassment (see Part II, the "Neuroscience has not been able to solve the 'hard problem'" section).

It led *Nature*, the most-cited scientific journal in the world, to flatly declare, "Despite a vast effort, researchers still don't understand how our brains produce consciousness."

By my fifties, then, based on my research, I knew that classical Newtonian science couldn't explain the strong evidence from NDE research (and other lines of evidence such as deathbed visions) that our consciousness survives death.

But when I tried to convince my good friend, a local university professor, I hit a brick wall.

The Professor was a committed Materialist with a hard-science background and a sharp mind. He and I went back a long way. When I came out with my first book, *Best Evidence,* someone suggested I give a copy to the professor. He read it carefully and responded with thoughtful counterarguments. He knew how to argue without turning it personal.

We discovered we both enjoyed a philosophical discussion and a large pepperoni pizza. Our amicable jousts became regular occurrences.

I expected the debate to be scrappy that day, and it was. We spent a Saturday afternoon that stretched into the early evening lolling on his back porch, enjoying a bottle of his favorite chardonnay, debating the survival of consciousness. The Professor wasn't buying any invisible afterlife state.

The intellectual tussle went back and forth. In a debate, he was a quiet listener but a deft counterpuncher. Against my NDE evidence, he cited the sheer weight of consensus science – a steady accumulation of discoveries and experiments over the last three centuries – which had concluded:

> (1) Only the physical exists. Only matter is real.
> (2) Consciousness exists, but it must therefore be made of matter.

(3) Consciousness is a product of the brain (the firing of neurons).
(4) The brain falls apart when we die.
(5) No brain, no consciousness.
(6) No consciousness, no afterlife.

The logic was inescapable, he declared. An afterlife was scientifically impossible. He finished off the bottle into my glass, then leaned back in his chair and returned to his. Yes, he conceded, the NDE research I cited provided some interesting counterevidence. And yes, he admitted, science had yet to explain exactly how the brain produced consciousness. But eventually it would, he assured me.

Meanwhile, a few puzzling scientific anomalies weren't enough to overturn a well-established scientific worldview. Tie goes to the defending champ. To get science to take an afterlife seriously, you had to prove Materialism wrong, knock it out of the ring. The Professor didn't see that happening.

Out on the lawn, night was falling. You could hear the crickets chirping in the shadows. He called his dog in, and corked the empty bottle. "The model we have works pretty well," he said. "I don't see any serious challenger waiting in the wings." His wife poked her head out the door. "You guys still talking? Dinner is getting cold." We got up and headed inside.

Another two years passed before I finally discovered that a powerful challenger had already stepped into the ring.

Quantum physics.

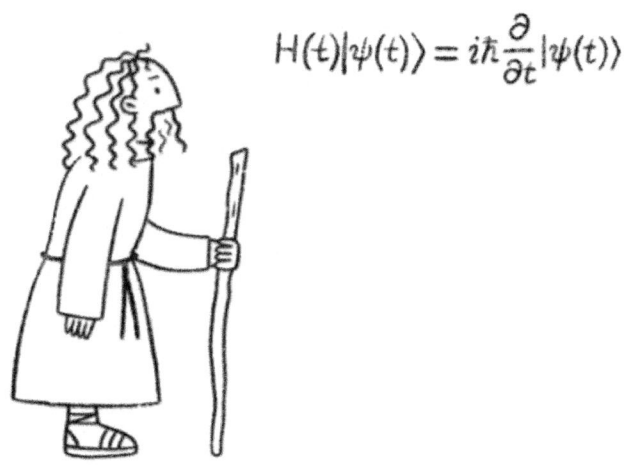

Quantum physics was first developed in the early 20th century, but the larger scientific community has ignored its full implications for almost a century, content to focus instead on the practical applications of classical Newtonian physics. Popular media infrequently reported on it, but my research for *Best Evidence* included monitoring frontier science.

I started noticing a number of magazine articles popping up on my Google science-news feed, discussing the weird but scientifically proven world of quantum physics, and how it differs so radically from the old 19th-century Newtonian physics.

They came from serious, credible publications – *Scientific American*, *Nature*, *New Scientist*, *Science Daily*, *Live Science*, *Science News*, *Discovery Magazine*, the *New York Times*, *Wired*, the *Atlantic*.

I started reading.

Quantum physics deals with a fundamental philosophical question – what is the nature of reality? It was a question I loved exploring as a philosophy major in college (it's a branch of philosophy called ontology).

I felt at home. I couldn't follow the dense math behind quantum physics, but I could easily follow the larger debate.

What I discovered in 21st-century quantum physics was a new, third view of reality – an updated physics without the 19th-century Newtonian baggage of scientific materialism.

Remove that unproven, philosophical assumption from science, and an afterlife is both scientifically possible and logical.

I had stumbled on ***quantum physics' hidden afterlife hypothesis – our consciousness is not made of matter.***

Warning: Science lesson ahead.

English polymath Sir Issac Newton (1642-1726) is one of the most brilliant and influential scientists in human history. He made groundbreaking discoveries and inventions in multiple fields, including physics, mathematics, optics, and astronomy. He invented calculus, built the world's first reflecting telescope, formulated the laws of gravity, and wrote the three laws of motion that explain how objects move and interact with each other. His amazing discoveries also sent science down a philosophical path that – since the discovery of quantum physics – increasingly looks like a dead end. It's called scientific materialism.

Throughout history, humans of every culture on earth have believed in the existence of a non-material, non-physical world that exists alongside the "natural" world of matter we experience with our five senses. That "spiritual" world included an afterlife state to which humans transitioned after physical death, their consciousness continuing.

Western science created the first systematic, sustained, successful challenge to that belief. Newton's insights and experiments – using the knowledge and tools available at the time – seemed to fully explain the nature and functioning of the universe as a world made of nothing but physical matter – tiny, hard particles that obeyed deterministic laws. There was no need to add an invisible dimension to the explanation – a God who created the world in seven days and caused floods and plagues to punish the wicked, or a heavenly realm in the sky. But you could still believe in God and an afterlife if you wanted, too (Newton himself apparently believed in both).

However, by the 19th century, many scientists, and a growing number of the educated public, had moved one step further.

Only the physical was real – things you could see, hear, taste, touch, smell, measure, and weigh. Science *required* you to reject anything not made of matter. That included both God and a non-physical dimension of reality called an afterlife.

Everyone seemed to forget that, while Newton's laws of motion and gravity were based on experimental proof and confirmed predictions, the additional conclusion that *only matter was real* wasn't science – it was simply an unproven philosophical assumption, no different than a religious belief.

Materialism (also known as Physicalism) became the religion of Newtonian science.

Its primary article of faith was simple: only things with size and weight, occupying a definite location in space, were real. These tiny, hard lumps of lifeless matter (atoms and the smaller particles of atoms) combine by pure chance to make larger and more complex "things" – rocks, molecules, cells, tissue, eyes, brains, bodies, humans, elephants, stars, galaxies, the universe.

But what do you do with consciousness?

Newtonian science accepted the reality of consciousness.

It had no choice. After all, Newton and his colleagues used consciousness (thinking and reasoning) to make their discoveries. But thoughts have none of the properties of matter. They have no size, no weight, don't occupy space. You can't see, hear, smell, taste, or touch a thought.

They certainly don't appear to be made of matter.

Confronted with this crisis of faith, Newtonian science initially simply ignored consciousness. Eventually, it settled on the belief that consciousness was a product of the brain, the firing of neurons. You take a bunch of neuron atoms, mix them together with electricity, and somehow you get the thought "Mary had a little lamb." Another specific combo of tiny bits of dead matter and electricity produces the thought "Maybe I'll have an espresso instead of a cappuccino." A third, specific combo might create the thought "This is absurd."

For the last hundred years, Newtonian science has struggled and failed to come up with an explanation for exactly how, where, or why mindless matter produces consciousness.

Quantum physics suggests it never will.

Quantum physics shatters the 19th-century Newtonian science belief that reality was made of matter. Quantum physicists discovered that the basic building blocks of our "physical" universe – atoms – are ultimately not even real "things."

As Werner Heisenberg, the Nobel laureate scientist credited with helping create quantum mechanics, explained: "The atoms or elementary particles themselves are not real; they form a world of potentialities or possibilities rather than one of things or facts."

Physical matter is fundamentally an illusion of our five senses.

Declared Heisenberg's fellow Nobel laureate Niels Bohr: "Everything we call real is made of things that cannot be regarded as real."

At the most fundamental level of reality, you don't find a world of tiny, hard, discrete particles obeying the laws of Newtonian physics. Matter – as understood by 19th-century science – disappears.

What doesn't disappear when reality turns out not to be made of matter? What continues to exist, unchanged, operating as usual?

Our consciousness. Our mind, observing this astonishing disappearance of matter.

Consciousness is fundamental, not matter.

According to Nobel Prize–winning quantum physics giant Erwin Schrödinger: "Consciousness cannot be accounted for in physical terms. For consciousness is absolutely fundamental."

Declares Nobel laureate physicist Brian Josephson: "Recognition that mind is fundamental rather than matter will be as significant a step for physics as the step from classical to quantum physics."

If reality is fundamentally not made of matter, then what is it made of?

Energy and consciousness.

Nobel Prize–winning physicist Max Planck, who coined the term "quantum" (the Latin word for "amount") and is credited with creating quantum theory, summarized his answer to the reality question in one, short, simple, now-famous sentence: "I regard matter as derivative from consciousness."

According to Planck, consciousness working at the most fundamental level of reality shapes packets ("quanta") of energy into the physical things that we experience with our physical five senses. Consciousness is primary; it exists first – then matter.

In a speech in Italy shortly before he died, Planck remarked: "As a man who has devoted his whole life to the most clearheaded science, to the study of matter, I can tell you as a result of my research about atoms this much: There is no matter as such. To me the term matter implies a bundle of energy which is given form by an intelligent spirit. All matter originates and exists only by virtue of a force which brings the particle of an atom to vibration and holds this most minute solar system of the atom together. We must assume behind this force the existence of a conscious and intelligent mind. Mind is the matrix of all matter."

Declared Planck's fellow Nobel laureate physicist Eugene Wigner, "Materialism was not consistent with present quantum mechanics." Materialism had to be abandoned for science to advance. "It was not possible to formulate the laws of quantum mechanics in a fully consistent way without reference to consciousness."

Quantum physics dethroned Materialism and made a place for non-material consciousness in science (*See Part II*). In doing so, it reopened a door to the scientific possibility of an afterlife.

Near-Death Experience research says our consciousness survives death. Quantum physics explains why this is scientifically possible and logical.

The logic is strong:

(1) There is *no scientific reason* today to assume that our consciousness is made of matter. Twenty-first century quantum physics has discovered that reality is fundamentally not made of Newtonian matter. Consciousness is thus not made of Newtonian matter.

(2) There is *no scientific proof* that our consciousness is a product of the brain. After a century of futile trying, Newtonian science still has not solved neuroscience's "hard problem."

(3) *The death and dissolution of the material brain doesn't extinguish consciousness because our consciousness is not made of matter, and it is not a product of the brain.*

(4) An afterlife is scientifically possible and logical.

Twenty-first-century quantum physics doesn't mean 19th-century Newtonian science is all wrong. It simply says its belief in the philosophy of Materialism was an unfortunate historical mistake.

Reality is fundamentally not made of Newtonian matter. Further, reality includes a non-physical consciousness that exists and operates both inside and outside of Newtonian space and time.

The possibility of an afterlife is not supernatural nonsense; it's simply modern science. So why do some scientists still refuse to even consider the possibility of an afterlife? My guess? Because letting go of scientific materialism would be professionally risky, intellectually demanding, and emotionally disturbing to them.

Quantum physics appears to validate what most of the world already accepts as true. According to a November 2019 *New Scientist* article, "Why Almost Everyone Believes in an Afterlife – Even Atheists," author Graham Lawton notes that most people hold curiously similar ideas about life after death. "Confronted with the finality of death, the majority of us, dogged rationalists included, cling on to the belief that it isn't the end. 'Most people believe in life after death,' says psychologist Jamin Halberstadt. . . . 'That's amazing. Science has changed the way we think about almost every aspect of our lives, including death, but through all of that, belief in life after death has remained steadfast.' Archaeological evidence for afterlife beliefs goes back at least 12,000 years, when bodies started to be buried with useful stuff to take to the other side. But such beliefs are far from a thing of the past. Surveys done regularly since the 1940s consistently show that about 70 per cent of U.S. citizens believe in some form of life after death – a number that is mirrored across the developed world. ...even the 30 per cent who say they don't, often do. When he asked extinctivists (people who totally reject the idea of life after death) whether they agreed with the statement 'conscious personality survives the death of the body, but I am completely unsure of what happens after that,' 80 per cent said yes."

This, says the author, suggests there is more to it than religion, fear, or an inability to imagine not existing.

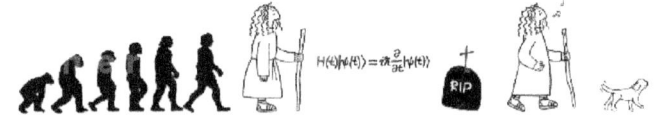

I could have stopped my quantum physics explorations there. I had uncovered a credible scientific hypothesis that supports the evidence from near-death experiences that our consciousness – yours and mine – can survive death. When my time comes, you can skip the flowers and do what you want with my mortal remains. I'm pretty confident my consciousness will be elsewhere, continuing my adventure. But instead of stopping there, I kept on reading – shifting my attention from science magazine articles to books like physicist Henry Stapp's well-argued *Mindful Universe: Quantum Mechanics and the Participating Observer*.

Why? Because I was fascinated by Max Planck's explanation of how consciousness creates the universe.

It's one thing to mentally accept the possibility of an afterlife – humans have believed this, and preached this, and taught this for millennia.

It's a completely different thing to mentally accept the possibility that your consciousness may actually help create the physical universe you experience with your five senses.

If it's true, then our importance and position in the universe soars. We humans move from disposable species to indispensable actors in the drama.

Let's take a second to step back a century.

Newtonian science says we humans are just accidental, by-products of a clockwork universe ruled by impersonal, mechanistic forces. Humans are nothing special. In the 19th century, we humans hit rock bottom in terms of how we saw our place in the universe.

We once viewed ourselves as creatures of God, ranking just below the angels, with a shot at heaven.

After the discoveries of Newton and Darwin, the scientific evidence seemed inescapable: we were simply intelligent monkeys, or meat robots – a transient, fleeting life-form destined to eventually be replaced through evolution with something better equipped to win an endless, grim survival-of-the-fittest contest. Materialist zoologist Richard Dawkins, in his book *The Selfish Gene*, famously and bluntly declared: "We are survival machines – robot vehicles blindly programmed to preserve the selfish molecules known as genes."

Believing in ourselves seemed embarrassingly unscientific.

Then quantum physics came along.

As we've learned, quantum physics has already produced a strong, logical, scientific argument for an afterlife. Can it also produce a credible scientific argument for restoring our belief in ourselves?

It already has.

According to the "consciousness causes collapse" variation of the current, leading interpretation of quantum physics (Copenhagen interpretation), our consciousness helps create the physical world we experience with our five senses. By the very act of our observing and measuring the subatomic quantum objects existing in a state of superposition, our consciousness collapses them into material "things" with size, weight, and location in physical space-time.

In other words, reality doesn't exist until consciousness observes it.

Wow. You might want to read that sentence again.

Leading backers of this explanation for how quantum physics works include celebrated mathematician John von Neumann, Nobel laureate physicist Eugene Wigner, and Wigner's Princeton colleague, physicist John Wheeler. The hypothesis is also backed by multiple experiments. Wheeler is famous for coining the term "black hole," and the concept of wormholes.

To Wheeler, we are not simply accidental bystanders on a cosmic stage; we are shapers and co-creators of the physical universe we experience. Per Wheeler, "The universe does not exist 'out there,' independent of us. We are inescapably involved in bringing about that which appears to be happening. We are not only observers. We are participators." (Wheeler's celebrated quote comes from Denis Brian's book *The Voice of Genius: Conversations with Nobel Scientists and Other Luminaries*).

Stay tuned, everybody. We may be in for an upgrade.

By now, I'm sure your head is spinning. I don't blame you. Quantum physics is a stunning, mind-blowing rewrite of everything we once believed about the ultimate nature of reality. Even Einstein struggled with it, so don't feel bad if your brain is hurting at this point in our lesson.

A *Science News* website feature called "Century of Science" sympathizes with our desire to cling to our old, comfortable, commonsense, 19th-century Materialist worldview. Quantum physics is psychologically wrenching, disorienting, bewildering.

"Quantum physics has drastically remodeled humankind's understanding of nature. In fact, a fair reading of history suggests that quantum theory is the most dramatic shift in science's conception of reality since the ancient Greeks deposed mythological explanations of natural phenomena in favor of logic and reason. After all, quantum physics itself seems to defy logic and reason. It doesn't, of course. Quantum theory represents the ultimate outcome of superior logical reasoning, arriving at truths that could never be discovered merely by observing the visible world."

But from the start of recorded history, we humans have always relied on observation of the visible world with our eyes (and other five senses) to decide what is real and what is not. "Seeing is believing," we say. Now science is telling us that the world of physical things we experience with our five senses is ultimately not real? *Science News* is damn right – it absolutely does "defy logic and reason."

We're used to thinking we live in a 100 percent physical universe of hard, concrete, physical "things." That's what Newtonian science has been telling us for almost 400 years now.

More importantly, that's what common sense and our physical eyes tell us.

You look around your dining room – there's a table with a Starbucks cup on it. The cup hasn't fallen through the table. You can touch the cup and hold it in your hand. If you draw in a deep breath, you can smell the coffee. Tipping it into your mouth, you can taste the coffee. It's delicious. Outside your house, you can see a maple tree and a fence. You can hear your clock ticking on the wall. It's 8 AM. When you leave your house to go to work, you slam the door, and you can hear it make a sound.

How can this physical world that you can see, taste, touch, smell, and hear every day, possibly not be what it appears to be to our eyes?

I spent months struggling to come up with a way to visualize in my head quantum physics' reversed view of reality – matter is not fundamental; consciousness is. I eventually remembered my high school English class. In his play, *As You Like It*, Shakespeare famously declared, "All the world's a stage."

OK, I said to myself, then how about this analogy?

Imagine you're sitting in a theater one evening, watching a play. You're entranced with the story because you're the playwright – you wrote the script, and it's based on your life. Up on the stage, an actor is sitting at a table with a coffee cup on it. He picks up the Starbucks, sniffs it deeply, then says, "This cappuccino smells absolutely delicious!" He tips it to his lips, then gives a satisfied sigh. He stares out the window at the maple tree and fence, then glances over at the clock on the wall. It shows 8 AM. According to the script, he leaves for work at this point in the play, ending the scene. He gets up and walks out, accidentally slamming the door.

Intermission. The house lights come up.

Squinting against the glare, you look around disoriented for a second. Then you suddenly remember that you have been watching a play. You were so entranced with the story, you forgot that all the "physical" props on the stage – the coffee, the room, the window, the fence, the tree, and even the time – are an illusion.

But everything certainly looked and felt real to you while you were in the moment.

As long as our consciousness is focused completely on the small, brightly lit stage (physical reality), and the rest of the theater (non-physical reality) remains in darkness, we will experience the props as real. It's only when the play ends (by physical death), and the lights are turned back on (when consciousness is no longer constrained by a physical brain built to only see physical reality), that we realize the illusion.

This "theater of the mind" analogy is not perfect, but it works for me.

Whew! OK, end of science lesson.

So what did I learn from my investigations? My biggest takeaway is this:

> *We've got strong scientific evidence from near-death-experiences research that our consciousness, our spirit, our soul – whatever we choose to call that which ultimately makes us be us – survives death.*
>
> *We've also got a modern, logical, scientific explanation for why that afterlife is possible. Our consciousness is not made of matter. The death and dissolution of the physical brain doesn't affect our non-material consciousness.*

You don't have to believe in God to believe in an afterlife. You can be agnostic about religion, or even an atheist, and still conclude – based on science alone – that an afterlife is possible.

Quantum physics doesn't say that God exists (or doesn't exist); it doesn't say that Christianity is better than any other religion (or worse); it doesn't support a specific afterlife condition – a Christian heaven with harps and angels, a Buddhist nirvana, or an American Indian great hunting ground in the sky.

Quantum physics simply says that there's no scientific reason that physical death should extinguish a non-material consciousness.

But this simple scientific conclusion can potentially have a profound effect on human society, once it seeps back into our collective psyche after centuries of scientific rejection and ridicule.

Science without spirit is not only bad science; it's toxic to humans. We need more than material comforts like food, clothing, and shelter to be happy, to thrive. We need hope, meaning, and purpose. In denying an afterlife, 19th-century Newtonian scientific materialism literally robbed humanity of its spirit; 21st-century quantum physics returns it to us.

Still skeptical?

Fine. But the debate rules have been rewritten. Rebuttals require science based on 21st-century quantum physics, not on blind faith in a scientifically discredited, 19th-century philosophy of Materialism.

Accepting the scientific possibility of an afterlife can change you.

Pioneering NDE researcher and psychiatrist Bruce Greyson notes that "near-death experiences seem to give people who have them the spark to reevaluate their lives and make changes in how they spend their time and how they relate to other people. They tell us that death is more about peace and light than about fear and suffering. They tell us that life is more about meaning and compassion than about wealth and control."

That kind of thinking can change the world.

Some other takeaways? Here are a few more:

- If we participate in bringing about the physical universe moment by moment at the quantum level with our consciousness – as the "consciousness causes collapse" hypothesis argues – then Newtonian science's "accidental" universe produced by blind chance is replaced by a universe imbued with human purpose and meaning. We're getting the world we believe in, we imagine, we choose, and we deserve.

- Quantum physics makes normal a whole range of "paranormal" phenomena I examined in *Best Evidence*. Multiple Nobel laureate quantum physicists have explored the relationship between quantum physics and phenomena like ESP, psychokinesis, synchronicity, and clairvoyance. A quantum physics phenomenon called "entanglement" may help explain how some of these psi phenomena work. They're worthy of scientific study.

- If consciousness can survive death, then it's logical to entertain the possibility that disembodied intelligences – e.g., spirits of the deceased, aka "ghosts," which humans of every culture on earth have reported for millennia – are real. If your mom says she saw grandma's ghost after she died, keep an open mind. It may have been grandma.

- The "consciousness is fundamental" hypothesis offers a way to view humans as more than just monkeys. Yes, our physical bodies and brains may be evolved from apes, but our consciousness is not, because it's not made of matter.

And the final lesson I learned on my journey?

Quantum physics is downright weird. My brain still hurts.

If you saw the wacky, wildly inventive 2022 movie *Everything Everywhere All at Once*, you got a cinematic visualization of two mind-blowing quantum physics discoveries – entanglement and superposition.

Australian quantum physicist James Quach takes you through the physics behind the script in a fun March 2023 article entitled *"Everything Everywhere All at Once: Quantum Physics Explained."*

Under the rules of quantum physics, a thing can be in two different places (here and there), or two different states (alive and dead) *at the same time*; effects can precede causes; time is an illusion; space is an illusion – two entangled particles separated by millions of light years can still remain instantly connected. It also suggests the possibility that, at the most fundamental level of reality, all is one – everything in the universe is part of a single, connected, unified whole.

Like it or hate it, we all need to get used to thinking in a new way about reality. Science has changed. We can't hide in the 19th century forever.

Quantum physics is currently science's most precise, powerful, proven theory of reality. It has successfully predicted countless experiments and spawned countless real-world applications in computing, cryptography, and electronics.

According to some estimates, roughly a quarter of our world's GDP relies on quantum mechanics.

But in the end, quantum physics' greatest benefit may turn out to be its positive effect on how we humans view our value and importance in the universe – and view our destiny when "what comes next" finally comes.

If religious belief no longer works for you when it comes to an afterlife, or if a nihilist 19th-century Newtonian-science worldview leaves you feeling small and worthless, you've now got a cheerful 21st-century-science alternative to carry you through life.

PART II: THE SCIENTIFIC EVIDENCE

Ready to dive into the scientific evidence? Here's how we'll do it.

In Part I, I make a series of claims. We're going to rewind the adventure back to the start and examine the scientific evidence for each. I'm a big Sherlock Holmes fan. He always walked around with a magnifying glass when hunting evidence. So each major claim starts with a magnifying glass.

During my investigation, I searched primarily for evidence that was current; that was backed by high-credibility scientists; and that appeared in an established science publication.

Science publications: My main targets included *Scientific American*, *Nature*, *New Scientist*, *Science News*, and other semi-technical publications. But scientific research focused on the revolutionary implications of quantum physics, the puzzling nature of consciousness, and hospital-based near-death experiences has exploded over the last two decades – along with coverage by the popular media. So I also kept an eye out for credible, well-written, insightful articles in respected general interest magazines and newspapers like *Wired*, the *Atlantic*, and the *New York Times*. I also came across fascinating TED Talks, YouTube interviews, and discussions on Reddit and Quora that broadened my understanding. When I cite an article, I give a title and date. That allows you to quickly find the article on the internet and read the whole thing to confirm that I am not selectively editing it. As much as possible, I have tried to avoid paraphrasing in favor of extended, direct quotes. Paraphrasing on my part increases the risk of misinterpretation (I'm a journalist, not a quantum physicist).

Current information: I targeted articles published as recently as possible. They include many articles published within the last five years. New discoveries in quantum physics appear regularly. For instance, until very recently, scientists believed that quantum effects were limited to the world of the infinitesimally tiny and that Newtonian classical physics still rules on the macroscopic scales of apples, humans, and stars. But in 2023, *New Scientist* magazine reported that researchers had succeeded in putting a sapphire crystal into a superposition of quantum states, bringing quantum effects into the macroscopic world. The implications of this experiment are huge: quantum physics applies to *all* reality, further pressuring us to abandon our old 19th-century Newtonian science worldview.

High-credibility scientists: When evaluating evidence and competing arguments, I gave extra weight to Nobel laureates. Nobel Prize winners make up the all-star team of science – the Michael Jordans, Tom Bradys, and Babe Ruths of their sport. It doesn't mean they're always right, but it does mean their opinions are backed by deep knowledge and experience; innovative, rigorous thinking; and the almost universal respect of their peers. With few exceptions, I also cite practicing researchers – working scientists active in the field – rather than celebrity science popularizers. Within the quantum physics field, I looked to quote quantum physicists rather than general scientists. Within the field of near-death-experience research, I favored medical researchers over anecdotal NDE reports. Many of the high-credibility scientists I quote also published books elaborating in depth on their ideas. I read many of them, looking to better understand their ideas and to ferret out interesting quotes to pass along. With one or two exceptions, the books I cite are accessible to the average, lay reader. If I can understand it, so can you.

OK, here we go.

NDE research strongly suggests that our consciousness survives death.

In 1975, a quiet, Georgia-born psychiatrist and physician, Dr. Raymond Moody, MD, the son of an agnostic surgeon, launched the modern, scientific study of near-death experiences (NDEs), forcing Newtonian science to take seriously the possibility of an afterlife.

NDEs have been around forever. Humans have been reporting NDEs for at least 2,000 years. One of the first NDE accounts comes from the Greek philosopher Plato (born 428 BC). In his book *The Republic,* Plato tells the story of a soldier named Er who awoke after being dead for twelve days to describe his account of his experience in the afterlife. But the rise of

Newtonian science caused a precipitous fall in interest in such stories. By the time Dr. Moody stumbled on NDE reports, science had buried the topic as unworthy of study.

Moody resurrected the topic from the dead with a small, thin book entitled *Life After Life*. It presented the personal stories of 100-plus ordinary people who had experienced a near-death experience (a term Moody coined). They were revived, and upon return to consciousness, they described what happened to them. The book went on to sell over 20 million copies.

Scientists could no longer simply dismiss them with a condescending wave of the hand and a simple "trust me, there's nothing to see here." They had to explain why literally millions of humans were reporting basically the same experience.

Once they were forced to look, scientists discovered that Moody's 100 NDE experiencers were just the tip of an iceberg. They discovered that NDEs are a universal human experience – 95 percent of the world's cultures report them, including our own. A 2011 study by the New York Academy of Sciences found that an estimated 9 million Americans have experienced an NDE (if you fall critically ill, you have a one-in-five chance of experiencing an NDE yourself).

In 1978, Moody and Bruce Greyson, MD, helped co-founded the International Association of Near Death Studies (IANDS). According to Greyson's research, near-death experiences are fairly common. Some 10 to 20 percent of people who come close to death report them. Researchers have identified the twelve elements most commonly found in an NDE (not every NDE experience includes all twelve):

- An out-of-body experience (OBE) involving separation of consciousness from the physical body.
- Heightened senses.
- Intense and generally positive emotions or feelings.
- Passing into or through a tunnel.

- Encountering a mystical or brilliant light.
- Encountering other beings (either mystical beings and/or deceased relatives or friends).
- A sense of alteration of time and space.
- A life review (a rewind of your life showing you how you affected everyone you met).
- Encountering an unworldly ("heavenly") realm.
- Encountering or learning special knowledge.
- Encountering a boundary or barrier.
- A return to the body, either voluntary or involuntary.

Moody's book, and the establishment of IANDS, allowed people who experienced an NDE to report their experiences without being ridiculed or treated as crazy. Today you can read several hundred books by people sharing their personal stories, from country guitar singers like Pam Reynolds Lowrey (see Part I), to hard-nosed scientists like Harvard Medical School neurosurgery professor Dr. Eben Alexander, who penned the 2012 *New York Times* bestseller, *Proof of Heaven: A Neurosurgeon's Journey into the Afterlife*. Alexander's years spent teaching neurosurgery at Harvard Medical School, and his deep, scientific grounding in neuroscience, helped his book find a huge audience among an increasingly open-minded public; and top NDE researchers, including Moody and IANDS experts, reviewed the book very favorably.

But die-hard skeptics argue that this mountain of anecdotal evidence isn't good enough – it's still theoretically possible that all 9 million NDE experiencers are either lying, deluded, or crazy – or reacting to drugs given during their medical crisis.

In the last twenty years, however, NDE researchers have conducted a number of scientifically designed, *hospital-based* studies that have put skeptics on the defensive, eliminating one skeptical hypothesis after another.

Peter Fenwick, MD, began researching NDEs in the United Kingdom in the 1980s. In his 1996 book, *The Truth in the Light: An Investigation of over 300 Near-Death Experiences*, he shared the results of his scientific analysis of more than 300 firsthand accounts submitted to him by NDE experiencers in the UK.

Fenwick specifically analyzes and addresses in detail skeptics' arguments that the NDE experience is due to the drugs given to patients as part of their resuscitation or to the effects of anesthesia, oxygen starvation, hypercarbia, hallucinations, endorphins, fear of death, or dissociation. Countering the drug-effect explanation, Fenwick notes that only 14 percent of his sample were given drugs of any kind at the time of their experience. So this explanation can be automatically ruled out for 86 percent of the cases.

Peter Fenwick, MD

How about oxygen starvation (anoxia) as causing the NDE experience? Fenwick's counter: Anoxia has been induced in experimental conditions by medical researchers thousands of times, in thousands of people; none of them have ever reported having a near-death experience. If anoxia is causing the NDE, why don't these thousands of people have an NDE?

Further, he notes, the characteristics of an NDE make anoxia improbable as an explanation. "As the brain becomes anoxic, it ceases to function. It becomes disrupted and disorganized, so that you become gradually confused, disoriented, your perceptions fragment, and finally you become unconscious. You do not think clearly, you do not have insights, you don't have clear, coherent visions. Yet one of the most fascinating things about the NDE is that the experience stands out as so clear and vivid."

Fenwick continues this analysis for twenty-seven pages, methodologically challenging, one by one, each of the remaining skeptical arguments.

"All [the skeptic's arguments] have significant drawbacks as a total explanation," Fenwick concludes. He ends with a last major question for skeptics to answer: "How is it that this coherent, highly structured [NDE] experience sometimes occurs during unconsciousness, when it is impossible to postulate an organized sequence of events in a disordered brain? One is forced to the conclusion that either science is missing a fundamental link which would explain how organized experiences can arise in a disordered brain, or that some forms of experience are transpersonal – that is, they depend on a mind which is not inextricably bound up with a brain."

Over on the Continent, Dutch cardiologist Pim van Lommel spent twenty years systematically studying, in the controlled environment of a cluster of hospitals with a medically trained staff, the near-death experiences of a wide variety of hospital patients who survived a cardiac arrest. In 2001, he and his fellow researchers published the study in the renowned medical journal *The Lancet*. The article caused an international sensation as it was the first scientifically rigorous study of the NDE phenomenon. In 2011, he

Pim van Lommel, MD

shared the results of his twenty-year scientific investigation in a book entitled *Consciousness Beyond Life: The Science of the Near-Death Experience*. Van Lommel's data from Holland extends and reinforces the rebuttals offered by Fenwick based on his UK study. He analyzes an even more exhaustive list of twenty-three specific skeptical explanations and explains why each fails as an argument.

Van Lommel spends thirteen pages challenging skeptics' *Physiological Arguments* – oxygen deficiency; carbon dioxide overload; chemical reactions in the brain including ketamine and endorphins;

psychedelics like DMT, LSD, psilocybin, and mescaline; electrical activity in the brain including epilepsy and artificial stimulation; and sleep disorders.

For example, Van Lommel notes that one of the first attempts at explaining away NDEs was based on the fact that stress releases endorphins. Maybe they explained the absence of pain and the sense of peace and well-being often reported by NDE experiencers? The problem, counters Van Lommel, is that the effect of endorphins usually lasts several hours, whereas the absence of pain and sense of peace during an NDE vanishes immediately after regaining consciousness (endorphins also fail to explain multiple other elements of NDEs).

Another example: Neuropsychologist Michael Persinger has famously carried out transcranial magnetic stimulation (TMS) experiments, which he believes resemble an NDE. But close examination of his findings disproves this, says Van Lommel. "The reported experiences, such as dreamlike, semi-mystical episodes with light and music, or the sense of somebody's presence, bear only a vague resemblance to the elements of an NDE. Suggestibility (that is, a placebo effect) seems to be the overriding factor in these reported experiences because Persinger also reports experiences in 33 percent of people without magnetic stimulation, and because a double-blind control of his research in Sweden failed to corroborate his results."

Van Lommel spends twelve pages analyzing the problems of skeptics' *Psychological Theories* – fear of death, expectations, depersonalization, dissociation, personality factors, fantasy and imagination, deceit, birth memories, hallucination, dreams, and delusion brought on by medication.

For example, some psychologists argue that fear of death causes NDE experiencers to imagine what they expect after dying. Yet Van Lommel notes that for many people the content of their NDE doesn't match their expectations. "Their experiences are identical, irrespective of whether they believe that death is the end of everything or whether they believe in life after death."

Further, children too young to have an idea of what to expect after death experience the same elements as adults.

How about hallucination as an explanation? That's been a favorite skeptical theory for decades. Sorry, says Van Lommel. "A hallucination is an observation without a basis in reality. The fact that an out-of-body experience during an NDE involves verifiable perceptions means that NDE is, by definition, not a hallucination."

Maybe NDE experiencers are mentally unstable? Sorry again, says Van Lommel. "Generally speaking, NDEs occur in mentally stable people who function normally in everyday life and who, except in age, do not differ from control groups without an NDE."

Is it all just fantasy-prone people with an overactive imagination? Easy to challenge, says Van Lommel. "The suggestion that an NDE is constructed on the basis of false memories or imagination can be refuted by the fact that people around the world report virtually identical NDEs.

Maybe the so-called NDE experiencer is telling a deliberate lie to look interesting or impress others? Skip the theorizing and simply talk with NDE experiencers, suggests Van Lommel.

"A personal meeting will quickly dispel such suspicions, not just because of what they say about the experience, but above all because of the obvious emotions and the struggle to find the right words when they share the experience. The fact the people often keep quiet for years for fear of rejection, and that when they finally talk about their NDE they do so only reluctantly to a handful of friends, also argues strongly against a deliberate lie to come across as interesting."

In the end, Van Lommel, like Fenwick, flags the same ultimate challenge skeptics face: how can you have lucid consciousness and verifiable perception during the loss or serious impairment of brain function? This seriously challenges the prevailing view of consciousness as a product of the brain. As a consequence, Van Lommel says, dogmatic scientists offer up a grab bag of one-sided, simplistic accounts of the NDE, resulting in theories that can sometimes account for one or two aspects on the NDE but never the whole NDE experience. Some skeptical theories are simply nonstarters. "Other theories start from unverified and unverifiable assumptions, or from speculation based on a few neurochemical studies of animal brains, which disqualifies them as a proper foundation for further debate."

The evidence is clear, Van Lommel concludes: "The theories on NDEs ... fail to explain the experience."

Back in the United States, radiation oncologist Jeffrey Long created in 1998 the Near-Death Experience Research Foundation (NDERF) and an online website that has collected over 1,300 NDE cases from around the globe – the largest NDE database in the world.

NDE experiencers of almost every race, creed, and culture answered a lengthy, professionally designed survey of 100-plus questions.

Jeffrey Long, MD

Long defines an NDE as an "event that takes place as a person is dying, or already clinically dead." They are considered near death "if they were so physically compromised that they would die if their physical condition did not improve." In addition, the experience had to be lucid – cases involving fragmented or disorganized memories were excluded from the study.

The experiencers were generally unconscious and often clinically dead, with an absence of heartbeat, breathing, or brain activity (when the heart stops beating, blood immediately stops flowing to the brain. Approximately ten to twenty seconds after that, the electroencephalogram, or EEG, which measures brain electrical activity, goes flat).

In 2010, he published a book entitled *Evidence of the Afterlife: The Science of Near-Death Experiences*. Long reviews a dozen specific skeptical objections to NDEs, including several not addressed by Fenwick or Van Lommel: maybe the NDE experiencer wasn't really near death? Or maybe the NDE experiencer's description of his experience was influenced by what the person read or saw about the NDE experience before the experience (the "Oprah factor")?

Then, turning the tables, Long challenges skeptics to explain away nine lines of evidence – based on a study of 600-plus cases from the NDERF database – that strongly suggest NDEs are real, and consciousness survives death. Long makes the following arguments:

- *"It is medically impossible to have a highly organized and lucid experience while unconscious or clinically dead."* Yet 74 percent of NDE experiencers report they had "more consciousness and alertness than normal," and an additional 20 percent reported experiencing "normal consciousness and alertness." Only 6 percent reported "less consciousness and alertness." A highly organized and lucid experience establishes that NDEs are not just dreams or hallucinations.

- "*NDE experiencers may see and hear in the out-of-body (OBE) state, and what they perceive is nearly always real.*" Approximately half of all NDE experiencers report an OBE. In his book, Long cites four major scientific studies of OBE experiencers' perceptions during resuscitation. The data showed extremely high levels of verified accuracy for the OBErs during their resuscitation, while control groups were highly inaccurate.

- "*NDEs take place among those who are blind, yet these NDEs often include visual experiences.*" For people born blind, sight is an abstract concept. Studies have shown that even their dreams don't include vision. Yet some born-blind NDE experiencers have reported full and clear vision.

- *"NDEs occur during general anesthesia when no form of consciousness should be taking place."* Yet it does. Long's study included twenty-three NDE experiencers reporting perception while under general anesthesia. Some 83 percent of them experienced "more consciousness and alertness than normal."

- *"A life review during the NDE accurately reflects real events in the NDE experiencer's life, even if those events have been forgotten."* In Long's study, eighty-eight NDE experiencers (14 percent) reported a life review. Long carefully reviewed those eighty-eight reports, asking the question: did either the NDE experiencer himself/herself or Dr. Long as an independent evaluator have any reason to doubt that any of the content of the scenes of (that person's) past life contained content was real? The conclusion: None of the eighty-eight life reviews contained unrealistic content. In fact, NDE experiencers having a life review were often impressed that their life review contained real details of their lives that they themselves had long forgotten.

- *"Virtually all beings encountered during NDEs are deceased at the time of the NDE, and most are deceased relatives."* Why would seeing deceased friends and relatives be evidence for life after death? Long argues that if NDEs were simply a product of brain function, one would expect them to see people recalled from recent memory, like the emergency personnel who helped them, or the person they had lunch with right before their heart attack. Instead, they see people they haven't seen or thought about in years or decades.

- *"The striking similarity of content in NDEs among very young children and that of adults strongly suggest that the content of NDEs is not due to preexisting beliefs."* Children five years of age or younger are practically a blank slate when it comes to preconceptions and belief about death. It's unlikely they have even heard about an NDE, much less have preconceptions of what it might include. Yet very young children in the Long NDE study reported experiencing every NDE element that older children and adults did. They have the same content as adult NDE experiencers.

- *"The remarkable consistency of NDEs around the world is evidence that NDEs are real events."* The NDERF files constitute the largest cross-cultural NDE database in the world. It includes NDE reports posted in non-English languages, with 250 volunteer translators worldwide helping to convert back and forth between twenty languages, including Arabic, Chinese, and Indonesian. The data confirm that the core NDE experience is the same all over the world. There are some minor differences, but they include the same elements, in basically the same frequency, and those elements appear to follow the same order of occurrence. The experience of dying appears similar among all humans.

- *"NDE experiencers are transformed in many ways by their experience."* NDE aftereffects often include transformations in the NDE experiencer's values, beliefs, and relations with others. Those aftereffects include a significant increase in a belief in an afterlife. People who personally experience an NDE are virtually unanimous in their conclusion that the afterlife is real.

Dr. Sam Parnia, MD, conducted his major NDE study between 2017 and 2020 on two continents. Some twenty-five hospitals in the United States and United Kingdom participated in the study, called AWARE II. Led by Parnia's researchers at New York University Grossman School of Medicine and elsewhere, the study involved 567 men and women whose hearts stopped beating while hospitalized and who received CPR between May 2017 and March 2020 in the

United States and United Kingdom. Its conclusions were summarized in a news release titled "Lucid Dying: Patients Recall Death Experiences During CPR," published by NYU Langone Health, New York University, on November 7, 2022. Bottom line? "Researchers found these experiences of death to be

Sam Parnia, MD

different from hallucinations, delusions, illusions, dreams, or CPR-induced consciousness." Here are a few additional key excerpts from the news release: "Survivors reported having unique lucid experiences, including a perception of separation from the body, observing events without pain or distress, and a meaningful evaluation of life, including of their actions, intentions, and thoughts toward others. A key finding was the discovery of spikes of brain activity, including so-called gamma, delta, theta, alpha, and beta waves up to an hour into CPR. Some of these brain waves normally occur when people are conscious and performing higher mental functions, including thinking, memory retrieval, and conscious perception."

Parnia penned the *New York Times* bestseller *Erasing Death: The Science That Is Rewriting Boundaries Between Life and Death*. Parnia argues that some form of afterlife may be uniquely ours, as evidenced by the continuation of the human mind and psyche after the brain stops functioning. In the final chapter, he concludes: "That entity that we define as consciousness, the soul, or the self – that which makes me who I am – does not stop existing just because someone has entered the period beyond death."

The scientific evidence for the postmortem survival of consciousness, produced by hospital-based NDE researchers like Fenwick, Van Lommel, and Parnia is enormous and compelling.

The last refuge of die-hard skeptics is to ignore this mountain of scientific evidence and hide behind the 19th century Newtonian science belief that only matter exists (therefore consciousness disappears forever with the death of the brain).

But in 1994, Australian David Chalmers blew a huge hole in the walls of their final fortress.

Neuroscience has not been able to solve the "hard problem."

Long-haired Australian-born philosopher David Chalmers coined the now famous term "the hard problem" facing Materialist neuroscientists who believe consciousness is a product of the brain. Today, Chalmers is professor of neural science at New York

University, as well as co-director of New York University's Center for Mind, Brain, and Consciousness. But he was a nobody when he first shocked the scientific establishment with his bold challenge.

Journalist Oliver Burkeman describes the drama of that first confrontation in his delightful article entitled "Why Can't the World's Greatest Minds Solve the Mystery of Consciousness?" published January 21, 2015, in the English newspaper, *The Guardian*:

"One spring morning in Tucson, Arizona, in 1994, an unknown philosopher named David Chalmers got up to give a talk on consciousness, by which he meant the feeling of being inside your head, looking out – or, to use the kind of language that might give a neuroscientist an aneurysm, of having a soul. Though he didn't realize it at the time, the young Australian academic was about to ignite a war between philosophers and scientists, by drawing attention to a central mystery of human life – perhaps the central mystery of human life – and revealing how embarrassingly far they were from solving it. . . . By the time Chalmers delivered his speech in Tucson, science had been vigorously attempting to ignore the problem of consciousness for a long time. . . . The scholars gathered at the University of Arizona – for what would later go down as a landmark conference on the subject – knew they were doing something edgy: in many quarters, consciousness was still taboo, too weird and new agey to take seriously, and some of the scientists in the audience were risking their reputations by attending. Yet the first two talks that day, before Chalmers's, hadn't proved thrilling. 'Quite honestly, they were totally unintelligible and boring – I had no idea what anyone was talking about,' recalled Stuart Hameroff, the Arizona professor responsible for the event. 'As the organiser, I'm looking around, and people are falling asleep, or getting restless.' He grew worried. 'But then the third talk, right before the coffee break – that was Dave.' With his long, straggly hair and fondness for all-body denim, the 27-year-old Chalmers looked like he'd got lost en route to a Metallica concert. 'He comes on stage, hair down to his butt, he's prancing around like Mick Jagger,' Hameroff said. 'But then he speaks. And that's when everyone wakes up.'"

Like the little boy in Hans Christian Andersen's famous folktale, *The Emperor's New Clothes*, Chalmers pointed out that 19th-century Newtonian science's explanation for consciousness figuratively had no clothes.

It was dressed in belief, not evidence.

Chalmers outlined the conundrum first in his seminal 1995 paper, "Facing Up to the Problem of Consciousness," and subsequently in his 1996 book, *The Conscious Mind: In Search of a Fundamental Theory.*

Chalmers points out that consciousness emerging from matter is not something one would logically predict from Newtonian physics. It makes no sense.

In his book, he writes: "Why should there be conscious experience at all? It is central to a subjective viewpoint, but from an objective viewpoint it is utterly unexpected. Taking the objective view, we can tell a story about how fields, waves, and particles in the spatio-temporal manifold interact in subtle ways, leading to the development of complex systems such as brains. In principle, there is no deep philosophical mystery in the fact that these systems can process information in complex ways, react to stimuli with sophisticated behavior, and even exhibit such complex capacities as learning, memory, and language.

"All this is impressive, but it is not metaphysically baffling.

"In contrast, the existence of conscious experience seems to be a new feature from this viewpoint. It is not something that one would have predicted from the other features alone. That is, consciousness is surprising. If all we knew about were the facts of physics, and even the facts about dynamics and information processing in complex systems, there would be no compelling reason to postulate the existence of conscious experience. If it were not for our direct evidence in the first-person case, the hypothesis would seem unwarranted; almost mystical, perhaps. Yet we know, directly, that there is conscious experience. The question is, how do we reconcile it with everything else we know?"

Materialism, Chalmers says, has no easy explanation as to how (or why) an electrochemical process in the physical brain – a three-pound bag of fat, water, protein, carbohydrates, salts, and electrical impulses – can create the personal, subjective qualities of consciousness.

"When we perceive, think, and act, there is a whir of causation and information processing, but this processing does not usually go on in the dark. There is also an internal aspect; there is something it feels like to be a cognitive agent. This internal aspect is conscious experience. Conscious experiences range from vivid color sensations to experiences of the faintest background aromas; from hard-edged pains to the elusive experience of thoughts on the tip of one's tongue; from mundane sounds and smells to the encompassing grandeur of musical experience; from the triviality of a nagging itch to the weight of a deep existential angst; from the specificity of the taste of peppermint to the generality of one's experience of selfhood. All these have a distinct, experienced quality.

"I know within myself that I exist. This is a common experience to every human being. But how do I know? I know because I feel in my interiority that I exist. The feeling carries the meaning (I exist). Therefore, the capacity to have feelings and understand their meaning is the essential property that 'explains' how we know. This is the crucial capacity of consciousness.

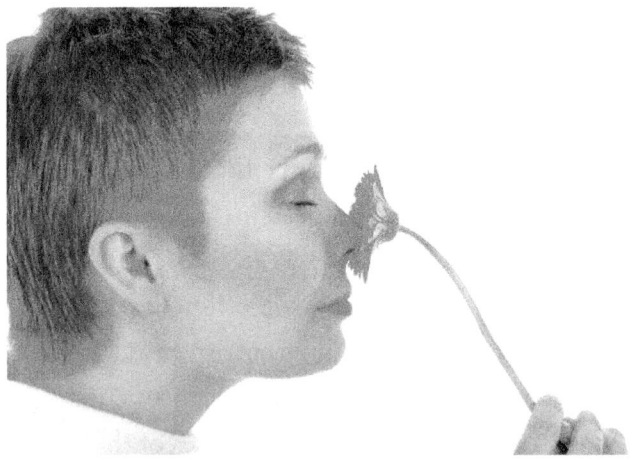

"When I smell a flower, I feel the scent. But the feeling is neither the set of electrical signals produced by the olfactory receptors inside my nose, nor the electrical signals generated by the brain after it has processed the olfactory signals. The output signals of the brain are correlated with the scent, but the scent is not equal to those signals. Electrical signals carry information, but that information is translated within my consciousness into a subjective feeling: the scent I feel within...

"By feeling the smell, seeing the image, and touching the petals of the jasmine, we connect with it in a special way. We 'experience' the jasmine, and this lived experience has special significance for us, producing other feelings like joy, curiosity, and love."

As Chalmers notes in his book, some scientists continue to insist that consciousness is an "illusion" but, Chalmers says, "I have little idea what this could even mean. It seems to me that we are surer of the existence of conscious experience than we are of anything else in the world." Instead, Chalmers declares, we need to move on from Materialism as an explanation for consciousness.

"Materialism is a beautiful and compelling view of the world, but to account for consciousness, we have to go beyond the resources it provides." Chalmers himself leans toward the idea that consciousness is both fundamental and possibly universal ("panpsychism" is an ancient idea that intrigues many modern scientists. See Joel Zadeh's delightful November 17, 2021, article in *Noema* magazine, entitled "The Conscious Universe").

Philosopher Chalmers defined the "hard problem" for the scientific community, but I personally found

Physicist Federico Faggin

physicist Federico Faggin's explanation a bit clearer and simpler to grasp. In his October 2020 article in *Meer* magazine, entitled "The Nature of Consciousness," Faggin starts off by defining consciousness as the capacity to have an inner experience based on sensations and feelings. These, he explains, are what philosophers call *qualia*. There are four basic categories of *qualia* – physical, emotional, mental, and spiritual:

(1) We feel physical pain.

(2) We feel emotions like love, hate, sympathy, curiosity, joy, sadness.

(3) We are self-aware; we think thoughts, plan, analyze, imagine, wish, reason, understand, choose.

(4) We have spiritual experiences of ecstasy, wonder, awe, existential despair.

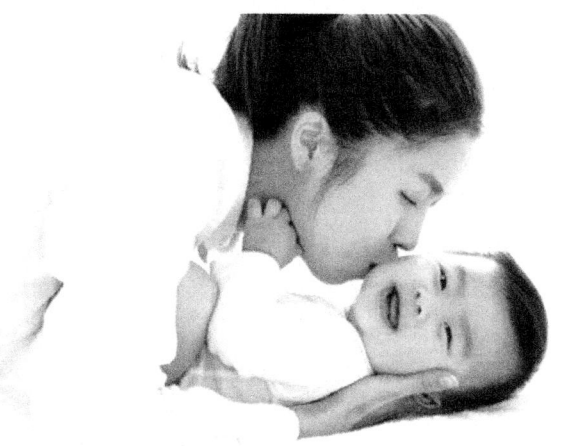

Atoms and electricity don't.

So how can combining things that don't have consciousness produce it?

The current neuroscience focus on correlations has gone nowhere, says science writer Dan Falk, in his September 25, 2023, article in *Scientific American*, entitled "Is Consciousness Part of the Fabric of the Universe?"

He writes: "Neuroscientists have identified a number of neural correlates of consciousness – brain states associated with specific mental states – but have not explained how matter forms mind in the first place."

MIT PhD graduate and University of California, Irvine, professor emeritus of cognitive sciences Donald Hoffman has won awards from both the National Academy of Sciences and the American Psychological Association for his research on perception, evolution, and consciousness. In a 2005 article in *Edge* magazine, entitled "What Do You Believe Is True Even Though You Can't Prove It?" Hoffman agrees with *Scientific American* writer Falk: Notes Hoffman, "Despite centuries of effort by the most brilliant of minds, there is as yet no physicalist theory of consciousness, no theory that explains how mindless matter or energy or fields could be, or cause, conscious experience. There are, of course, many proposals for where to find such a theory – perhaps in information, complexity, neurobiology, neural Darwinism, discriminative mechanisms, quantum effects, or functional organization. But no proposal remotely approaches the minimal standards for a scientific theory: quantitative precision and novel prediction."

Hoffman himself has concluded that consciousness is fundamental and has developed a formal model of consciousness based on a mathematical structure called conscious agents. In a fascinating YouTube interview on *The Tim Ferriss Show*, entitled "Does Consciousness Survive Death?" Hoffman says: "If consciousness is fundamental ... there is the possibility that some aspect of my consciousness survives death." He then goes on to explain how his own model of consciousness might be scientifically tested.

Hoffman's thought-provoking scientific explorations have appeared in *Scientific American* and *New Scientist*, and his March 2015 TED Talk, titled "Do We See Reality as It Is?" has received almost 4 million views.

As we learned earlier, Dr. Sam Parnia was chief investigator at New York University on the 567-patient AWARE II NDE study. In his book *Erasing Death*, Parnia cites four specific problems with the classic Newtonian materialist belief that you can "make mind from meat" – i.e., that mind is a product of the physical brain:

(1) The theory "does not provide a plausible mechanism that may account for the development of consciousness, thoughts, and all that makes up the human psyche or soul from brain cell activity. The theories simply propose potential intermediary pathways that

may be mediating consciousness. But correlation is not causation."

(2) "The nature and substance of thought (the amalgamation of which comprises the self) seems inherently different from electricity, or a chemical, or any protein-based substance we know. Most (classical) scientific theories simply say thought exists, and it comes about from the brain, but can't specify how, where, or why."

(3) "How do occurrences that are preconscious (in other words, chemical or electrical events that are continuously going on in the brain but are not part of our 'conscious awareness' such as the effect of hormones or other events that are instead occurring in the unconscious domain) become conscious, other than to say that it somehow does occur at a critical point?"

(4) "We know that a fundamental part of our lives involves the notion of free will. We are judged in society based upon our intentions and actions, and the brain-based views . . . cannot account for this. If correct, they would mean that our lives would be completely determined by our genes and environment and hence there would be no place for personal accountability."

Parnia's first argument – there's no explanation for exactly *how* the brain produces consciousness – is echoed by Bruce Greyson, professor of psychiatry and neurobehavioral sciences at the University of Virginia, who helped co-found the International Association of Near Death Studies (IANDS): "The dirty secret of neuroscience is that we have no idea how a physical event like electrical current or a chemical change in a nerve cell can produce consciousness."

Parnia's second argument – the radical difference between "meat and mind" – is highlighted by Dutch philosopher-scientist Bernardo Kastrup. Kastrup earned his PhD in computer engineering with specializations in artificial intelligence and reconfigurable computing; and helped design and program computer experiments for the Large Hadron Collider, the world's largest and most powerful particle accelerator. In his provocatively titled book *Why Materialism is Baloney,* Kastrup offers up a clear, detailed, and comprehensive attack on Materialism in general; while simultaneously driving home Parnia's argument that mind and matter are radically different. Declares Kastrup, "There is no way to logically deduce the qualities of conscious perception, cognition, or feeling from the mass, momentum, spin, position, or charge of the subatomic particles making up the brain."

Neuroscientist Mario Beauregard raises the same point in his book *Brain Wars*: "The brain can be weighed, measured, scanned and dissected, and studied. The mind that we conceive to be generated by the brain, however, remains a mystery. [Mind] has no mass, no volume, no shape, and it cannot be measured in space and time. Yet it is as real as neurons, neurotransmitters, and synaptic junctions."

Beauregard notes that neuroscience methods allow researchers to measure physical and chemical correlates of mental events, not the mental events themselves.

"Correlations between mental activity and brain activity do not imply causation and identity.... The mistaken belief that mental events are identical with their neural correlates is still entertained by some science writers, journalists and neuroscientists, and it leads to what has been called 'the neurological fallacy': the erroneous attribution of mental properties to parts of the brain (or to the brain itself)." Beauregard points to the extensive, groundbreaking research performed by neurosurgeon Wilder Penfield. Penfield performed brain mapping on over a thousand patients over the course of several decades. "At the end of his scientific career, Penfield concluded that higher mental functions – such as consciousness, reasoning, imagination, and will – are not produced by the brain; mind is a non-physical phenomenon interacting with the brain." Contemporary materialists, Beauregard notes, argue that sooner or later neuroscience will be able to completely explain mind and consciousness. "These materialists do not seem to realize that future technological development will only allow neuroscientists to measure more refined correlates of mental activity."

So what is the relationship between the brain and consciousness? Beauregard argues that the findings of contemporary near-death-experience research suggests the mind acts like a radio transmitter, and the brain like a radio receiver. Turning off the receiver (the brain) doesn't stop the transmitter (the mind) from continuing to transmit. "Equating 'mind' with 'brain' is as illogical as listening to music on a radio, demolishing the radio, and thereby concluding that the radio was creating the music," he writes. NDE studies confirm this: consciousness can continue to operate even when the brain is completely shut down (i.e., during clinical death).

The brain as radio receiver is one way to view the brain.

A second way is to view the brain as simply a filter of consciousness.

Psychiatrist Bruce Greyson uses this analogy in his book *After: A Doctor Explores What Near-Death Experiences Reveal About Life and Beyond*: "The brain is a device for the mind to act more effectively on the physical body, to focus our thoughts on the physical world.... The brain works like a filter to block out information that the body doesn't need for survival, and selects from the thoughts and memories stored in them only the information the body needs." This isn't a new idea, says Greyson. He cites the writing of philosophers, thinkers, and scientists from ancient Greece right up to modern times.

His fellow neuropsychiatrist and NDE researcher Peter Fenwick hammers home the same argument in a *Psychology Today* interview published June 26, 2019, entitled "Can Consciousness Exist Outside of the Brain?"

The author notes that "The prevailing consensus in neuroscience is that consciousness is an emergent property of the brain and its metabolism. When the brain dies, the mind and consciousness of the being to whom that brain belonged cease to exist. In other words, without a brain, there can be no consciousness.

"But according to the decades-long research of Dr. Peter Fenwick, a highly regarded neuropsychiatrist who has been studying the human brain, consciousness, and the phenomenon of near-death experience (NDE) for 50 years, this view is incorrect. Despite initially being highly incredulous of NDEs and related phenomena, Fenwick now believes his extensive research suggests that consciousness persists after death. In fact, Fenwick believes that consciousness actually exists independently and outside of the brain as an inherent property of the universe itself like dark matter and dark energy or gravity.

"Hence, in Fenwick's view, the brain does not create or produce consciousness; rather, it filters it.

"As odd as this idea might seem at first, there are some analogies that bring the concept into sharper focus. For example, the eye filters and interprets only a very small sliver of the electromagnetic spectrum, and the ear registers only a narrow range of sonic frequencies. Similarly, according to Fenwick, the brain filters and perceives only a tiny part of the cosmos' intrinsic 'consciousness.'

"Indeed, the eye can see only the wavelengths of electromagnetic energy that correspond to visible light. But the entire EM spectrum is vast and extends from extremely low energy, long-wavelength radio waves to incredibly energetic, ultrashort-wavelength gamma rays. So, while we can't actually 'see' much of the EM spectrum, we know things like X-rays, infrared radiation, and microwaves exist because we have instruments for detecting them.

"When the eye dies, the electromagnetic spectrum does not vanish or cease to be; it's just that the eye is no longer viable and therefore can no longer filter, be stimulated by, and react to light energy. But the energy it previously interacted with remains nonetheless. And so too when the ear dies, or stops transducing sound waves, the energies that the living ear normally responds to still exist.

"According to Fenwick, so it is with consciousness. Just because the organ that filters, perceives, and interprets it dies does not mean the phenomenon itself ceases to exist. It only ceases to be in the now-dead brain but continues to exist independently of the brain as an external property of the universe itself. . . .

"So, ironically, only in death can we be fully conscious."

Kastrup offers up additional scientific evidence for the brain simply being a filter of consciousness.

Kastrup makes his arguments in a March 29, 2017, article in *Scientific American*, entitled "Transcending the Brain." The article flags the odd scientific fact that, in some unusual cases of physical damage to the brain, the actual result is *enriched* consciousness or cognitive skill. He cites reliable reports in the medical literature of bullet wounds to the head, stroke, concussion, meningitis, and even the progression of dementia leading to *expanded* cognitive and artistic skills.

He includes the famous case of stroke victim Jill Taylor. In December 1996, the 37-year-old Harvard-trained brain scientist experienced a massive stroke in the left hemisphere of her brain. During her stroke, she observed her mind deteriorate to the point that she could not walk, talk, read, write, or recall any of her life – all within four hours.

At the same time that her brain was severely impaired, her consciousness expanded: "My perception of my physical boundaries was no longer limited to where my skin met air. I felt like a genie liberated from its bottle. The energy of my spirit seemed to flow like a great whale gliding through a sea of silent euphoria."

Her experience is detailed in her *New York Times* best-selling book, *My Stroke of Insight: A Brain Scientist's Personal Journey.*

How is it that by damaging or disabling the physical brain you can actually expand consciousness? The simplest explanation is that the brain doesn't produce consciousness; it limits it. This allows our non-local, non-physical consciousness to operate in Newtonian, physical space and time.

Dr. Bruce Greyson, in his book *After,* cites one final argument supporting the "brain-as-filter" hypothesis which resonated with me personally. Greyson noted that a rare brain phenomenon called "terminal lucidity" involves sudden, unexplainable, clear thinking in advanced Alzheimer's patients. The fact that it happens at all is a complete puzzle for neuroscientists. Greyson explains: "This astonishing and unexplained recovery usually happens in the hours before the person dies, suggesting that the deteriorating brain has lost its ability to filter the mind, which is briefly free to express itself before the person dies."

It is amazing, indeed, as my sister Jody, an emergency room nurse in Connecticut, will vouch to you. Our mother, Rayanna, suffered from severe Alzheimer's in the last few years of her life. By 2002, she had been in a nursing home for two years. She no longer recognized any of her family, including her six kids or our spouses or children. When I visited her from Hawaii that spring, she stared blankly at me, her son, with no clue as to who I was. But the morning she died, a week before Christmas, lying in her Danbury hospital bed surrounded by Jody and my brother-in-law Bill, she suddenly opened her eyes and addressed them by name, conducting a lucid conversation, despite her verified, years-long, full Alzheimer's condition.

Returning to Parnia's four arguments against mind as a product of the brain, let's finish with his objection that it doesn't recognize or allow for free will. Materialism says only matter exists; humans are only made of matter; if so, we're subject to the same deterministic, causal laws that apply to everything else in the physical universe. Like falling rocks, humans don't have the freedom to choose between paths A or B; they simply follow fixed, immutable laws of nature.

Thus, materialists believe in a philosophy of Determinism – that the laws of physics fully explain all that we do and that everything that happens is fixed in advance, predetermined by what happened before in a long causal chain. Under this theory, ethics and morality is nonsense. If I rob, or torture, or kill you, I had no choice. I'm just a meat robot following the fixed laws of nature. Obviously, society would collapse if it embraced this 19th-century corollary of Materialism. But quantum physics has overturned Determinism as well as Materialism.

The first blow came in 1927, with the proof of Heisenberg's "Uncertainty Principle" of quantum physics. It says we cannot know both the position and the speed of a particle with perfect accuracy. As Heisenberg noted, "The more precisely we determine the position, the more imprecise is the determination of velocity." Author William Egginton, in his book *The Rigor of Angels: Borges, Heisenberg, Kant, and the*

Ultimate Nature of Reality, notes that this discovery demolished classical science's belief in Determinism. "Since Newton had exposited the laws of motion, scientists had accepted the notion, proposed most famously by the French mathematician Pierre-Simon Laplace, that perfect knowledge of an object's position and momentum, and the forces acting on it in the present, will yield perfect knowledge of its future permutations. Heisenberg's discovery put this determinism to rest."

In 2006, Princeton University mathematician John Conway, the John von Neumann Professor in Applied and Computational Mathematics, and his colleague Simon Kochen, confirmed Determinism's demise. They published, in *Foundations of Physics,* their now famous "Free Will Theorem." For the backstory of how they came up with their proof, google the March 23, 2009, article by Kitta MacPherson, entitled "High-Powered Mathematicians Take on Free Will." MacPherson explains that by testing to see whether the behavior of particles is predetermined, they ended up tackling the centuries-old debate over the existence of free will. "'It's not about theories anymore – it's about what the universe does,' said Kochen. And we've found that, from moment to moment, nature doesn't know what it's going to do. A particle has a choice.'"

As the decades pass, the "hard problem" continues without a materialist solution.

On June 24, 2023, the prestigious science journal *Nature* published the results of the latest failure by scientific materialists to solve Chalmers' "hard problem." Twenty-five years earlier, in 1998,

neuroscientist Christof Koch bet philosopher David Chalmers a case of wine that the mechanism by which the brain's neurons produce consciousness would be discovered by 2023. The *Nature* article's title summarized the results of the bet: "Decades-Long Bet on Consciousness Ends – and It's Philosopher 1, Neuroscientist 0." Concluded the author: "Despite a vast effort, researchers still don't understand how our brains produce (consciousness)." The *New York Times* agreed, stating flatly: "No one has found a clear neural correlate of consciousness."

Maybe it's time for neuroscience to finally drop its 19th century science belief that Newtonian matter is fundamental.

Twenty-first century science has proven it's not.

Reality is ultimately not made of matter.

Is there any hard scientific evidence that matter isn't fundamental – that reality, at its most fundamental level, is not made of tiny, indivisible, hard lumps of dead matter obeying Newtonian laws, like we all learned in high school science?

There is.

Are you ready for some serious science?

Meet Dartmouth College physics professor Marcelo Gleiser. In a January 8, 2023, article in *Big Think*, entitled "Quantum Mystery: Do Things Only Exist Once We Interact with Them?" he takes us step by step down the rabbit hole: of quantum physics.

"Perhaps the weirdest thing about the quantum world is that the notion of an 'object' falls apart. Outside the world of molecules, atoms, and elementary particles, we have a very clear picture of an object as a thing we can behold. This applies to a door, a car, a planet, and a grain of sand. Moving to smaller things, the concept still holds for a cell, a virus, and a large biomolecule like DNA.

"But it is here, at the level of molecules and of distances shorter than a billionth of a meter or so, that the problems begin. If we keep moving to smaller and smaller distances, and continue to ask what are the objects that exist, quantum physics kicks in. 'Things' become fuzzy, their shapes unclear and their boundaries uncertain.

"Objects evaporate into clouds, as elusive in their contours as words are to describe them. We can still think of crystals as being made of atoms arranged in certain patterns – like our familiar kitchen salt, which is made of cubic lattices of sodium and chlorine atoms. But dive into the atoms themselves, and simple pictures evaporate in a puff of puzzlement."

Newtonian matter vanishes.

Astrophysicist Adam Frank is a good "go-to" guy for puzzled journalists (like me) seeking lay-friendly scientific explanations. He is a regular commentator on NPR's *All Things Considered*, and won the American Physical Society's 2020 Burton Award for his "multi-channel promotion of public understanding of physics."

When I first came across Werner Heisenberg's shocking statement about matter (atoms and particles) not being real, I realized I needed help to wrap my head around the idea. I found in Frank's March 13, 2017, article in *Aeon*, entitled "Minding Matter", a clear, accessible explanation of "matter" – and its implications for the debate over whether consciousness is a product of the brain.

Exactly what is "matter"? Frank asks. "After more than a century of profound explorations into the subatomic world, our best theory for how matter behaves still tells us very little about what matter is. Materialists appeal to physics to explain the mind, but in modern physics the particles that make up a brain remain, in many ways, as mysterious as consciousness itself."

Nineteenth-century Newtonian science viewed the world as made up of particles – tiny, hard, concrete physical "things" with definite properties like mass and volume and a location in space. And those particles always acted in a very predictable way.

Frank writes, "Back in the good old days of Newtonian physics, the behaviour of particles was determined by a straightforward mathematical law that reads $F = ma$. You applied a force F to a particle of mass m, and the particle moved with acceleration a. It was easy to picture this in your head. Particle? Check. Force? Check. Acceleration? Yup. Off you go. The equation $F = ma$ gave you two things that matter most to the Newtonian picture of the world: a particle's location and its velocity. This is what physicists call a particle's state. Newton's laws gave you the particle's state for any time and to any precision you need. If the state of every particle is described by such a simple equation, and if large systems are just big combinations of particles, then the whole world should behave in a fully predictable way. Many materialists still carry the baggage of that old classical picture."

But, Frank says, in the early 20th century, Albert Einstein and Max Planck introduced the idea of the quantum, sweeping away the old classical view of reality.

"In Isaac Newton's physics, position and velocity were indeed clearly defined and clearly imagined properties of a particle. Measurements of the particle's state changed nothing in principle. The equation $F = ma$ was true whether you were looking at the particle or not.

"All of that fell apart as scientists began probing at the scale of atoms early last century. In a burst of creativity, physicists devised a new set of rules known as quantum mechanics. A critical piece of the new physics was embodied in Schrödinger's equation. Like Newton's $F = ma$, the Schrödinger equation represents mathematical machinery for doing physics; it describes how the state of a particle is changing. But to account for all the new phenomena physicists were finding (ones Newton knew nothing about), the Austrian physicist Erwin Schrödinger had to formulate a very different kind of equation.

"When calculations are done with the Schrödinger equation, what's left is not the Newtonian state of exact position and velocity. Instead, you get what is called the wave function (physicists refer to it as psi after the Greek symbol Ψ used to denote it). Unlike the Newtonian state, which can be clearly imagined in a commonsense way, the wave function is an epistemological and ontological mess. The wave function does not give you a specific measurement of location and velocity for a particle; it gives you only probabilities at the root level of reality. Psi appears to tell you that, at any moment, the particle has many positions and many velocities. In effect, the bits of matter from Newtonian physics are smeared out into sets of potentials or possibilities.

"It's not just position and velocity that get smeared out. The wave function treats all properties of the particle (electric charge, energy, spin, etc.) the same way. They all become probabilities holding many possible values at the same time.

"Taken at face value, it's as if the particle doesn't have definite properties at all. This is what the German physicist Werner Heisenberg, one of the founders of quantum mechanics, meant when he advised people not to think of atoms as 'things.' Even at this basic level, the quantum perspective adds a lot of blur to any materialist convictions of what the world is built from."

To our physical five senses, matter appears to be a solid object, with a definite size, shape, and weight, occupying a definite location in space. But it's an illusion of our five senses.

When we drill down to the fundamental level of reality, we find that Newtonian matter disappears, replaced by packets of energy (called "quanta" from the Latin plural word for "amount"; singular: "quantum").

These energy packets, says Heisenberg, are "something standing in the middle between the idea of an event and the actual event, a strange kind of physical reality just in the middle between possibility and reality."

Sometimes these packets of energy are called quantum "particles." But a quantum "particle" and a Newtonian particle are *fundamentally different* in their properties and the principles that govern their behavior. Let's count the ways: Unlike Newtonian matter, these packets of energy: 1) inherently have the properties of both a particle and a wave *at the same time* (Wave-Particle Duality); 2) cannot be measured precisely (the Uncertainty Principle); 3) before they're observed, exist in multiple places or states of being *at the same time* (Superposition); 4) can be entangled with other quantum objects, so that the state of one can instantaneously affect the state of another, regardless of distance (Non-Locality); 5) their behavior can be affected by *simply observing them* (Observer Effect); and 6) don't follow the deterministic laws of 19th-century

Newtonian physics (Quantum Indeterminacy). In short, they are not made of Newtonian matter, Newtonian matter simply does not exist at the most fundamental level of reality.

What does this mean for the Materialist argument that our consciousness is made of matter? It means neuroscience must look elsewhere for a solution to the "hard problem."

The logic is strong: *if reality is ultimately not made of Newtonian matter, then consciousness is not made of Newtonian matter.*

Declares Frank: "Our current understanding of matter alone is unlikely to explain the nature of mind."

Quantum physics says the 19th-century Newtonian philosophy of Materialism is false.

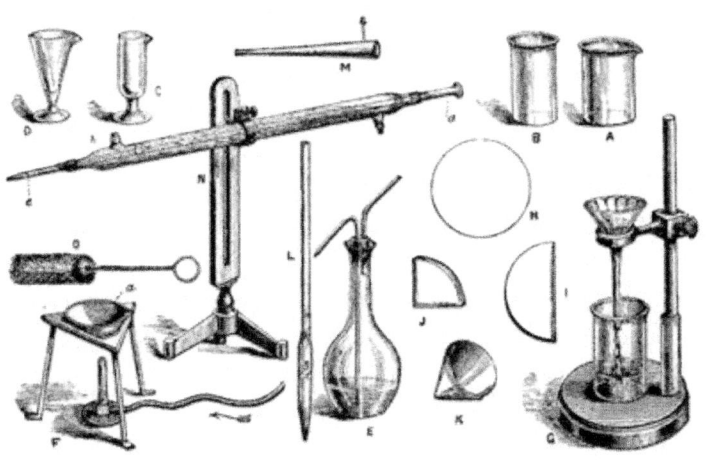

19th-century scientific tools

If Materialism is false, an afterlife is possible.

Remember that long, back-and-forth debate I had with the Professor that evening over a bottle of chardonnay? He made a strong logical, 19th-century argument:

(1) If only Newtonian matter is real, then our consciousness must also be made of matter.
(2) Matter falls apart when we die.
(3) Therefore an afterlife is impossible.

But we no longer live in the 19th century.

Heisenberg, Schrodinger, and 21st-century quantum physics have conclusively demonstrated that reality is fundamentally *not* made of Newtonian matter.

If so, then the Professor's afterlife argument collapses. His compelling logic instead gives the bottle of wine to me:

If reality is fundamentally not made of Newtonian matter, then consciousness is logically not made of Newtonian matter. If consciousness is not made of Newtonian matter, then the death and dissolution of the material brain does not affect a non-material consciousness. Thus *the survival of consciousness is scientifically both logical and possible.* This is quantum physics' hidden afterlife hypothesis.

Ready for a pop quiz? (I never had the chance to tease my great friend, the Professor, with this test; he passed away before I wrote this book. But I'm convinced he's smiling along, somewhere in the afterlife state).

Below is a "Who's Who" list of six Nobel laureates in quantum physics and their enormous contributions to the creation and development of quantum physics:

- *Max Planck* originated quantum theory.
- *Werner Heisenberg* created quantum mechanics.
- *Niels Bohr* teamed with Heisenberg to create the current, leading explanation of quantum mechanics, called the Copenhagen Interpretation.
- *Eugene Wigner* earned a Nobel Prize for his discoveries related to the atomic nucleus and elementary particles.
- *Erwin Schrödinger* created the foundational equation that describes the evolution of the wave function of a quantum system. Schrödinger was also the first person to explore the phenomenon of quantum entanglement – a term he coined, and a quantum physics property at the heart of today's quantum technologies, including quantum computers, quantum communications, and quantum sensors.
- *Brian Josephson* invented the "Josephson junction" circuit, which today is found hardwired into everything from quantum computer prototypes to instruments that measure neural activity inside the human brain.

Question: What else do these six have in common besides their Nobel Prizes?

Answer: All six rejected Materialism.

Nobel Prize–winning quantum physicist Werner Heisenberg, who helped pioneer quantum mechanics, believed the history of science could be divided into two periods, based on what scientists thought about matter. In a series of lectures in the 1950s, Heisenberg argued that, at the beginning of the 20th century, we entered a new period – quantum physics threw off the philosophy of Materialism that dominated the natural sciences of the 19th century. Regarding Materialism, Heisenberg wrote: "[This] frame was so narrow and rigid that it was difficult to find a place in it for many concepts of our language that had always belonged to its very substance, for instance, the concept of mind, of the human soul, or of life. Mind could be introduced into the general picture only as a kind of mirror of the material world."

Classical physics' belief in Materialism was unscientific, Heisenberg declared. "In classical physics, science started from the belief – or should one say, from the illusion – that we could describe the world, or least parts of the world, without any reference to ourselves."

Like Church clerics in 1610, faced with Galileo's celestial discoveries, many scientists refused to give up their old-world view. Noted Heisenberg: "The violent reaction to the recent development of modern physics can only be understood when one realizes that here the foundations of physics have started moving; and that this motion has caused the feeling that the ground would be cut from under science."

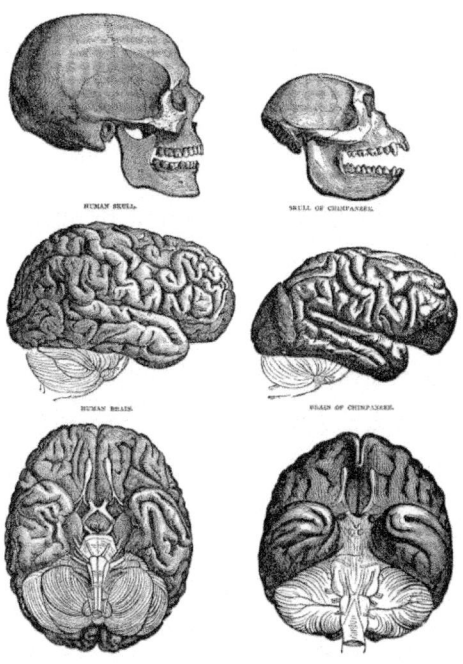

19th-century study of human vs. monkey brain

Neuroscientist Mario Beauregard, in his book *Brain Wars*, gives credit to materialist science's role in advancing knowledge but says quantum physics dealt scientific Materialism a fatal blow. "Materialist science, based on the classical Newtonian physics, took science out of the Dark Ages, showing us a world no one had ever seen before. Now there is another heretofore invisible world for us to see, one that the dogmas of materialist science obscure, but that is brought into focus by the discoveries of quantum physics. Towards the end of the 19th century, it became obvious that classical physics was limited; it was just not able to explain certain phenomena at the atomic level. The acknowledgement of these limitations led to the development of a revolutionary new branch of physics called quantum mechanics (QM) which smashed the scientific materialist world view."

A recent, literal example of the smashing of matter occurred on July 4, 2012, when the Large Hadron Collider, operated by the European Organization for Nuclear Research (CERN), announced that its particle-smashing experiments had uncovered evidence for the Higgs boson. This "particle," called by some the "God particle," and the Higgs quantum field that causes it, have been hypothesized as being necessary to explain how mass comes into being.

As science writer Jim Baggott says in his recent book, *Higgs: The Invention and Discovery of the God Particle*, "Mass is constructed entirely from the energy of interactions involving naturally massless elementary particles.... The physicists kept dividing, and in the end found nothing at all."

Philosopher Graham Smetham flags the implications of the experiment: "Mass, and so matter, are derived aspects of an insubstantial process of reality. In fact, they're also products of aspects of reality that are immaterial, i.e. not material at all." Smetham elaborates further in his article "On 'Known-to-Be-False' Materialist Philosophies of Mind." As he notes, "However, this conclusion should not come as a shock, for if there is one thing that has been established by the science of quantum mechanics, it is the fact that 'materialism' must be abandoned as a viable metaphysical position."

Theoretical physicist Marcelo Gleiser, of Dartmouth University, and astrophysics professor Adam Frank of the University of Rochester, argue that Materialism faces two intractable problems. They highlight them in their article "The Blind Spot," published in the magazine *Aeon* on January 8, 2019:

"Behind the Blind Spot sits the belief that physical reality has absolute primacy in human knowledge, a view that can be called scientific materialism. In philosophical terms, it combines scientific objectivism (science tells us about the real, mind-independent world) and physicalism (science tells us that physical reality is all there is). Elementary particles, moments in time, genes, the brain – all these things are assumed to be fundamentally real. By contrast, experience, awareness, and consciousness are taken to be secondary. The scientific task becomes about figuring out how to reduce them to something physical, such as the behaviour of neural networks, the architecture

Physicist Marcelo Gleiser

of computational systems, or some measure of information.

"This framework faces two intractable problems.

"The first concerns scientific objectivism. We never encounter physical reality outside of our observations of it. Elementary particles, time, gene, and the brain are manifest to us only through our measurements, models, and manipulations. Their presence is always based on scientific investigations, which occur only in the field of our experience.... These tests never give us nature as it is in itself, outside our ways of seeing and acting on things. Experience is just as fundamental to scientific knowledge as the physical reality it reveals.

"The second problem concerns physicalism. According to the most reductive version of physicalism, science tells us that everything, including life, the mind, and consciousness, can be reduced to the behaviour of the smallest material constituents. You're nothing but your neurons, and your neurons are nothing but little bits of matter. Here, life and the mind are gone, and only lifeless matter exists. To put it bluntly, the claim that there's nothing but physical reality is either false or empty. If 'physical reality' means reality as physics describes it, then the assertion that only physical phenomena exist is false. Why?

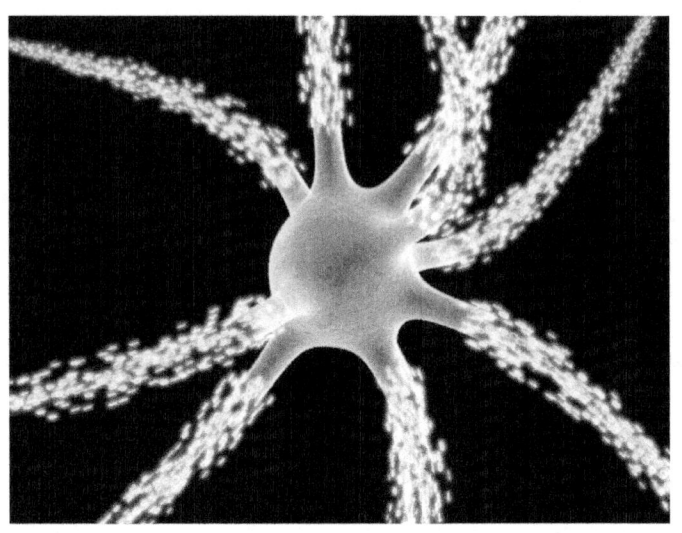

Brain neuron

Because physical science – including biology and computational neuroscience – doesn't include an account of consciousness. This is not to say that consciousness is something unnatural or supernatural. The point is that physical science doesn't include an account of experience; but we know that experience exists, so the claim that the only things that exist are what physical science tells us is false. On the other hand, if 'physical reality' means reality according to some future and complete physics, then the claim that there is nothing else but physical reality is empty, because we have no idea what such a future physics will look like, especially in relation to consciousness.

"Objectivism and physicalism are philosophical ideas, not scientific ones – even if some scientists espouse them. They don't logically follow from what science tells us about the physical world, or from the scientific method itself. By forgetting that these perspectives are a philosophical bias, not a mere data-point, scientific materialists ignore the ways that immediate experience and the world can never be separated."

Frank and Gleiser have since expanded their *Aeon* magazine article into a full book. Theoretical physicist turned neuroscientist Alex Gomez-Marin reviewed *The Blind Spot* in the February 29, 2024, issue of the journal *Science*, citing the book's "potential to become a classic text." Joined by philosopher Evan Thompson, physicists Frank and Gleiser demonstrate that the blind spot (the scientific materialism assumption) is hidden in plain sight in every branch of science — including neuroscience.

Asking how the brain gives rise to consciousness begs the question of whether it actually does so. "[The authors'] conclusion is unflinching: 'the hard problem [of consciousness] is an artifact of the Blind Spot.'" Drop the unrecognized and unproven scientific materialism assumption, and the hard problem turns easy — because consciousness is not a product of the brain.

The emergence of quantum physics is forcing all branches of science to justify or abandon their unproven, 19th century Materialist assumptions. Some are struggling.

In a *Scientific American* article entitled "Is Scientific Materialism 'Almost Certainly False?'" published January 30, 2013, science writer John Horgan highlights the growing unease within the scientific community regarding Materialism (and other long-held assumptions of Newtonian science):

"Science's limits have never been more glaringly apparent.... Fields such as neuroscience, evolutionary psychology and behavioral genetics and complexity have fallen far short of their hype. Some scholars, notably philosopher Thomas Nagel, are so unimpressed with science that they are challenging its fundamental assumptions. In his new book *Mind and Cosmos: Why the Materialist Neo-Darwinian Conception of Nature Is Almost Certainly False*, Nagel contends that current scientific theories and methods can't account for the emergence of life in general and one bipedal, big-brained species in particular. To solve these problems, Nagel asserts, science needs 'a major conceptual revolution,' as radical as those precipitated by heliocentrism, evolution and relativity.

"Many pundits calling for such a revolution are peddling some sort of religious agenda, whether Christian or New Age. Nagel is an atheist, who cannot accept God as a final answer, and yet he echoes some theological critiques of science.

"'Physic-chemical reductionism,' he writes, 'cannot tell us how matter became animate on Earth more than three billion years ago; nor can it account for the emergence in our ancestors of consciousness, reason, and morality.' Evolutionary psychologists invoke natural selection to explain humanity's remarkable attributes, but only in a hand-wavy, retrospective fashion, according to Nagel. A genuine theory of everything, he suggests, should make sense of the extraordinary fact that the universe 'is waking up and becoming aware of itself.' In other words, the theory should show that life, mind, morality, and reason were not only possible but even inevitable, latent in the cosmos from its explosive inception. Nagel admits he has no idea what form such a theory would take; his goal is to point out how far current science is from achieving it."

It's taken the broader scientific community, steeped in the philosophy of Materialism, decades to catch up with the implications of quantum physics. Like all scientific revolutions, the paradigm shift takes time.

But we're seeing some shift in attitude, and it appears to be accelerating in scientific circles.

In 2014, a movement in the scientific community, called a Campaign for Open Science, issued an eighteen-point declaration called "Manifesto for a Post-Materialist Science."

It has gathered the signatures of more than 500 scientists, professors, MDs, and thought leaders.

The first six points of the manifesto expose the materialist assumptions embedded in classical Newtonian physics, and call for a return to dogma-free science:

(1) "The modern scientific worldview is predominantly predicated on assumptions that are closely associated with classical physics. Materialism – the idea that matter is the only reality – is one of these assumptions. A related assumption is reductionism, the notion that complex things can be understood by reducing them to the interactions of their parts, or to simpler or more fundamental things such as tiny, material particles.

(2) "During the 19th century, these assumptions narrowed, turned into dogmas, and coalesced into an ideological belief system that came to be known as 'scientific materialism.' This belief system implies that the mind is nothing but the physical activity of the brain, and that our thoughts cannot have any effect upon our brains and bodies, our actions, and the physical world.

(3) "The ideology of scientific materialism became dominant in academia during the 20th century. So dominant that a majority of scientists started to believe that it was based on established empirical evidence, and represented the only rational view of the world.

(4) "Scientific methods based upon materialistic philosophy have been highly successful in not only increasing our understanding of nature but also in bringing greater control and freedom through advances in technology.

(5) "However, the nearly absolute dominance of Materialism in the academic world has seriously constricted the sciences and hampered the development of the scientific study of mind and spirituality. Faith in this ideology, as an exclusive explanatory framework for reality, has compelled scientists to neglect the subjective dimension of human experience. This has led to a severely distorted and impoverished understanding of ourselves and our place in nature.

(6) "Science is first and foremost a non-dogmatic, open-minded method of acquiring knowledge about nature through the observation, experimental investigation, and theoretical explanation of phenomena. Its methodology is not synonymous with Materialism and should not be committed to any particular beliefs, dogmas, or ideologies."

You can add 21st century quantum physics to the list of endorsers of a Post-Materialist Science worldview. Quantum physics is literally "post-materialist science" – it declares that Newtonian matter is not even real.

As we have learned, this scientific discovery is what makes an afterlife scientifically possible and logical. Beyond restoring the possibility of an afterlife, does 21st-century, post-Materialist, quantum physics offer any other cheerful, potential surprises for us humans? Here's one more for you.

How about the possibility that the human mind has a hidden superpower?

Reality doesn't exist until consciousness observes it.

Does human consciousness create the physical universe we experience with our five senses? It's a bold hypothesis, but it's championed by some notable quantum physics heavyweights, and supported by multiple scientific experiments.

Author Michael Brooks, in his 2012 article in *New Scientist* entitled "Reality: How Does Consciousness Fit In?" writes: "How do many possibilities become one physical reality? This is the central question in quantum mechanics, and has spawned a plethora of proposals, or interpretations. The most popular is the

Copenhagen interpretation, which says nothing is real until it is observed, or measured."

Celebrated physicist John von Neumann (who pioneered the formal mathematical foundation of quantum mechanics), Nobel laureate Eugene Wigner, and Wigner's Princeton colleague John Wheeler are among the giants of quantum physicists who argue that the act of measurement ultimately requires a conscious observer. A mechanical detector alone is not enough. (Wheeler pioneered the theory of nuclear fission with Niels Bohr. He also first introduced the concept of wormholes, and coined the term "black hole").

Physicist and Dartmouth professor Marcelo Gleiser - explains their argument in his January 2018 blog post "Consciousness, Quantum Physics, and Reality":

"Very serious scientists such as Nobel laureate Eugene Wigner and his Princeton colleague John Wheeler have considered the role of consciousness in physics and to what point it determines the reality in which we live. When we measure something as small as the smallest bits of matter, we must use a detector. We don't have direct sensorial contact with an electron or with an atom. Their existence is registered when they interact with (the electrons and atoms of) a detector and we hear a click or see a pointer move.

"In the orthodox interpretation of quantum physics, known as the Copenhagen Interpretation, it is this interaction that determines the existence of the particle: before the measurement we can't even say that the particle exists!

"In other words, according to this way of thinking about quantum physics, something only exists after it is detected. Wigner and physicist John Wheeler suggested that without an observer to set things up and interpret the results, without a consciousness with intent, this measurement wouldn't make sense. This being the case, the particle's existence is contingent on its interaction with the human consciousness.

"More dramatically and in italics, *consciousness has an active role in determining what exists*."

Wigner's fellow Nobel laureate Werner Heisenberg included a conscious observer in his explanation of how physical things come into existence.

"The discontinuous change in the wave function takes place with the act of registration of the result by the mind of the observer. It is this discontinuous change of our knowledge in the instant of registration that has its image in the discontinuous change of the probability function."

Distinguished British-American physicist Freeman Dyson, celebrated for his work in quantum field theory, echoes Heisenberg:

Physicist Freeman Dyson

"At the level of single atoms and electrons, the mind of an observer is involved in the description of events. Our consciousness forces the molecular complexes to make choices between one quantum state and another."

For a closer look at the "consciousness causes quantum collapse" hypothesis – how consciousness transforms the quantum world of potentialities and possibilities into one of physical things or facts – I suggest Kelvin McQueen's 2017 article "Does Consciousness Cause Quantum Collapse?"

His conclusion: "The 'consciousness causes quantum collapse' hypothesis – when combined with modern neuroscience – is a viable theory of physical and mental reality, which offers a clear research program and distinctive experimental predictions. It proposes a solution to the measurement problem by defining when and where collapse occurs. And it provides a place for consciousness in nature by giving consciousness a causal role."

McQueen and David Chalmers, in their May 2021 paper "Consciousness and the Collapse of the Wave Function," review three major arguments against the "consciousness causes collapse" hypothesis and argue that advanced versions of the hypothesis remain compatible with empirical evidence. More importantly, in principle, these versions of the theory can be tested by experiments with quantum computers.

The creation of physical reality includes all of us, working together, suggests Robert Lanza and his physicist co-authors Dmitriy Podolskiy and Andrei Barvinsky – theorists, respectively, in quantum gravity and quantum cosmology. Lanza is the author of *Biocentrism: How Life and Consciousness are the Keys to Understanding the True Nature of the Universe*. The trio summarize their scientific hypothesis in a June 7, 2021,

Big Think interview, entitled "Is Human Consciousness Creating Reality?"

As the article author notes, "Lanza contends that a network of observers is necessary and is 'inherent to the structure of reality.' As he explains, observers – you, me, and anyone else – live in a quantum gravitational universe and come up with 'a globally agreed-upon cognitive model' of reality by exchanging information about the properties of spacetime. 'For, once you measure something,' Lanza writes, 'the wave of probability to measure the same value of the already probed physical quantity becomes 'localized' or simply collapses. That's how reality comes to be consistently real to us all. Once you keep measuring a quantity over and over, knowing the result of the first measurement, you will see the outcome to be the same. Similarly, if you learn from somebody about the outcomes of their measurements of a physical quantity, your measurements and those of other observers influence each other – freezing the reality according to that consensus.' Adds Lanza, 'a consensus of different opinions regarding the structure of reality defines its very form, shaping the underlying quantum foam.'"

Can Nature by itself, or an inanimate detection machine, replace human consciousness in terms of performing the act of observation/measurement, as some Materialist-minded physicists believe?

Stanford University physicist Andrei Linde argues no. He lays out his argument in a *Discovery* magazine article published May 31, 2002, entitled "Does the Universe Exist if We're Not Looking at It?" Linde says conscious observers are an essential component of the universe and cannot be replaced as observers by inanimate objects.

"The universe and the observer exist as a pair," Linde contends. "You can say that the universe is there only when there is an observer who can say, Yes, I see the universe there. These small words – it looks like it was here – for practical purposes it may not matter much, but for me as a human being, I do not know any sense in which I could claim that the universe is here in the absence of observers.

"We are together, the universe and us. The moment you say that the universe exists without any observers, I cannot make any sense out of that. I cannot imagine a consistent theory of everything that ignores consciousness. A recording device cannot play the role of an observer, because who will read what is written on this recording device? In order for us to see that something happens, and say to one another that something happens, you need to have a universe, you need to have a recording device, and you need to have us.

"It's not enough for the information to be stored somewhere, completely inaccessible to anybody. It's necessary for somebody to look at it. You need an observer who looks at the universe. In the absence of observers, our universe is dead."

Distinguished quantum physicist Henry Stapp agrees with Linde. He lays out his arguments in a May 29, 2018, article in *Scientific American,* entitled "Coming to Grips with the Implications of Quantum Mechanics." Stapp's credibility within quantum physics circles is impeccable. Stapp received his PhD in particle physics at the University of California, Berkeley, under the supervision of two Nobel laureates – Emilio Segrè and Owen Chamberlain. He subsequently was recruited by, and worked closely with, two more Nobel laureate giants in quantum physics – Werner Heisenberg and Wolfgang Pauli. Later in his career, he worked with Princeton physicist John Wheeler on problems in the foundations of quantum mechanics area.

Stapp argues that the logic is irrefutable: the output of a detector only becomes known when it is consciously observed.

"The bottom line is that we cannot know that detectors actually perform measurements, for we cannot abstract ourselves out of our knowledge . . . the output of a detector only becomes known when it is *consciously observed* by a person. Recall Max Planck's position: 'I regard consciousness as fundamental. I regard matter as derivative from consciousness. *We cannot get behind consciousness.*' (Emphasis added.) . . . *only conscious observers* can perform measurements."

Like Stapp and Linde, Wheeler argued that a machine, or a mechanical detector of some sort, wasn't enough to change quantum possibilities into physical things. We humans are required.

As Wheeler famously declared, "The universe does not exist 'out there,' independent of us. We participate in bringing the universe into reality."

Wheeler expanded on his hypothesis in a fascinating interview published in *Futurism* magazine on February 3, 2014, entitled "John Wheeler's Participatory Universe." Here are some thought-provoking excerpts from the article:

"In the final decades of his life, the question that intrigued Wheeler most was: 'Are life and mind irrelevant to the structure of the universe, or are they central to it?' He suggested that the nature of reality

Physicist John Wheeler

was revealed by the bizarre laws of quantum mechanics. According to the quantum theory, before the observation is made, a subatomic particle exists in several states, called a superposition (or, as Wheeler called it, a 'smoky dragon'). Once the particle is observed, it instantaneously collapses into a single position. Wheeler suggested that reality is created by observers and that: 'no phenomenon is a real phenomenon until it is an observed phenomenon.' He coined the term 'Participatory Anthropic Principle' (PAP) from the Greek 'anthropos', or human. He went further to suggest that 'we are participants in bringing into being not only the near and here, but the far away and long ago.'

"This claim was considered rather outlandish until his thought experiment, known as the 'delayed-choice experiment,' was tested in a laboratory in 1984. This experiment was a variation on the famous 'double-slit experiment' in which the dual nature of light was exposed (depending on how the experiment was measured and observed, the light behaved like a particle (a photon) or like a wave. Unlike the original 'double-slit experiment,' in Wheeler's version, the method of detection was changed AFTER a photon had passed the double slit. The experiment showed that the path of the photon was not fixed until the physicists made their measurements.

"[The results] proved what Wheeler had always suspected – observers' consciousness is required to bring the universe into existence."

(The results of this groundbreaking,1984 experiment were reconfirmed in a famous, follow-up experiment published in *Science* in October, 2007, entitled " Wheeler's delayed-choice thought experiment: Experimental realization and theoretical analysis." It was reconfirmed yet a third time in 2015 by physicists at the Australian National University (*Nature Physics*, May 25, 2015, "Wheeler's delayed-choice gedanken experiment with a single atom.")

Consciousness is fundamental, not matter.

Why science ignored the implications of quantum physics for a century.

Abandoning the 19th-century Newtonian physics assumption that only matter is real solves not only the "hard problem" in neuroscience. It also allows us to make sense of the many intellectually counterintuitive findings of quantum physics – but only if scientists are willing to look.

Physicist Sean Carroll laments a surprising timidity within the scientific community in terms of its willingness to explore the full implications of quantum physics in his September 7, 2019, article in the *New York Times*, entitled "Even Physicists Don't Understand Quantum Physics: Worse, They Don't Seem to Want to Understand It."

As he notes, "Few modern physics departments have researchers working to understand the foundations of quantum theory. On the contrary, students who demonstrate an interest in the topic are gently but firmly – maybe not so gently – steered away, sometimes with an admonishment to 'Shut up and calculate!'" Carroll senses the situation is gradually changing, and a new generation of philosophers of physics are doing important work in bringing conceptual clarity to the field. But Carroll says science needs to step up its efforts here. "After almost a century of pretending that understanding quantum mechanics isn't a crucial task for physicists, we need to take this challenge seriously."

Why don't more scientists take up the challenge? Because they're afraid of being accused of adopting an unscientific, "mystical" worldview, argues Alan Steinberg, in his September 2021 article in *Psychology Today* entitled "Why Physicists Don't Want to Understand Quantum Mechanics."

"Why would physicists not want to understand science's most successful theory and furthermore, why would they discourage their students from trying? Based on the early history of quantum mechanics, it is my theory that the answer is they already knew where such an understanding would lead. It leads to the 'mystical' worldview held by eastern Vedic philosophy, and other western philosophies, that everything *is* ultimately consciousness. . . .

"If we assume that 'mystical' worldview, then Max Planck's statement 'I regard consciousness as fundamental. I regard matter as derivative from consciousness' becomes logical. 'Subject and object are only one' now makes sense because the actual subject of any observation is consciousness, and we are assuming the object is made of consciousness. And, the fact that our universe does not have separability, that everything is instantaneously connected, now is no longer 'spooky' because if everything is ultimately consciousness then there is only one thing in our universe, consciousness. Everything is instantly connected because everything is ultimately the same thing, consciousness.

"And, in assuming that everything is consciousness we make Chalmers's hard problem of consciousness easy. If everything is ultimately consciousness, then subjective experience is primary and fundamental. We do not need to explain how brains can produce consciousness because we are assuming brains do not produce consciousness. Brains are made of consciousness."

I suspect some scientists also strongly resist abandoning their 19th-century Newtonian science worldview because they fear doing so could allow religion to slip in the back door, as the professor wrote me after reading *Best Evidence*. As long as

consciousness is viewed as made of matter – a product of the brain – that can't happen. But quantum physics suggests that consciousness is fundamental, not matter. Religion could claim that this "consciousness" is just another word for "soul."

Is it? Not necessarily.

True, the soul, like consciousness, survives death. But people who talk about a "soul" usually mean a thing given to you by God, who created you, and tells you what to believe and how to live. That soul can be "lost" or "saved" by how you ethically live.

"Consciousness" is a scientific term. It just refers to our internal, personal, subjective experiences – self-awareness (I exist), thoughts, feelings, memories. It doesn't necessarily imply the existence of a God, or a belief in any specific religious dogma. You can be an atheist and still conclude – based on the scientific evidence alone – that consciousness is fundamental, is not made of matter, and thus can survive death.

But some scientists would argue it introduces confusion. Why go looking for trouble? Better to ignore the full implications of quantum physics, and keep the door shut.

Yet an increasing number of scientists are embracing the risk and opening the door – some out of curiosity, and others out of frustration with the current, creaky, 19th-century Newtonian explanation of consciousness.

Curiosity drives astrophysicist Paul Sutter. In his thoughtful December 2022 article in *Live Science*, entitled "Does Consciousness Explain Quantum Mechanics?" he writes: "One of the most perplexing aspects of quantum mechanics is that tiny subatomic particles don't seem to 'choose' a state until an outside observer measures it. The act of measurement converts all the vague possibilities of what could happen into a definite, concrete outcome. While the mathematics of quantum mechanics provides rules for how that process works, that math doesn't really explain what that means in practical terms.

"One idea is that consciousness – an awareness of our own selves and the impact we have on our surroundings – plays a key role in measurement and that it's our experience of the universe that converts it from merely imagined to truly real. But if this is the case, then is it possible that human consciousness could explain some of the weirdness of quantum mechanics? . . . The standard interpretation of quantum mechanics, known as the Copenhagen interpretation, says to ignore all this and just focus on getting results. In that view, the subatomic world is fundamentally inscrutable, and people shouldn't try to develop coherent pictures of what's going on. Instead, scientists should count themselves lucky that at least they can make predictions using the equations of quantum mechanics.

"But to many people, that's not satisfying. It seems that there's something incredibly special about the process of measurement that appears only in quantum theory. This specialness becomes even more striking when you compare measurement to, say, literally any other interaction. . . .

"Because consciousness is so important to humans, we tend to think there is something special about it. After all, animals are the only known conscious entities to inhabit the universe. And one way to interpret the rules of quantum mechanics is to follow the above logic to its extreme end: What we call a measurement is really the intervention of a conscious agent in a chain of otherwise mundane subatomic interactions."

For physicist Federico Faggin, frustration drives his exploration of the "consciousness is fundamental" hypothesis. The intellectually curious Faggin helped pioneer the technology underlying nearly all contemporary integrated circuits; designed the world's first microprocessor, the Intel 4004; researched artificial neural networks; and developed the early touchpads and touchscreens now used in nearly all mobile devices. He is the recipient of many honors and awards, including the National Medal of Technology and Innovation from President Barack Obama in 2010.

In his article in *Meer* magazine from December 2020, entitled "Consciousness Is Fundamental," Faggin argues: "Physics assumes the existence of matter, energy, space, and time (MEST) with certain properties and seeks to derive all other observable properties by using a mathematical theory based on relationships between those fundamental properties. These relationships define the basic laws of physics that are held to be universally valid and immutable.

"The soundness of this approach is predicated on the experimentally verifiable predictions made by the theory when applied to any phenomenon. The basic assumptions, however, are considered true without proof. And herein lies a problem, because the existence of our conscious inner reality is impossible to explain with the current 'matter-first' theories.

"Consciousness cannot arise from matter devoid of it any more than electricity and magnetism could emerge from elementary particles devoid of electrical charge and magnetic spin. The current 'scientific' explanation that consciousness arises from complex organizations of unconscious matter is completely inadequate because complexity has nothing to do with consciousness. The only reasonable possibility to make progress is to *postulate* that consciousness is an irreducible property of nature."

Physicist John Stewart Bell was nominated for the Nobel Prize and would have won it if he hadn't died prematurely. Bell's Theorem proved the reality of quantum entanglement – the "spooky action at a distance" that Albert Einstein fought against and lost. He foresaw a similar ending to the debate over whether science should bring consciousness into their model of reality. "As regards mind, I am fully convinced that it has a central place in the ultimate nature of reality."

Consciousness is not the only thing 19th-century Newtonian science deliberately refused to examine.

It also refused to look at evidence for anything it felt violated its belief that only physical matter existed. If something couldn't be weighed, measured, and perceived by the five senses, it wasn't real. The only "scientific" explanation acceptable to 19th-century Newtonian science was hoax, delusion, or gullibility.

That included not only God, heaven, and ghosts; it also included common psychic phenomena persistently reported by humans for millennia — extra sensory perception (telepathy, precognition, clairvoyance), and psychokinesis (the ability of the mind to influence matter). Each of these psi phenomena provides additional evidence that the mind is not bound by 19th-century Newtonian physics "laws," and is not a product of the physical brain.

Newtonian science branded these phenomena "pseudoscience," and attacked any scientist curious enough to even dare look at the evidence.

But Nobel laureates don't scare easily. They've got the credentials and credibility to withstand ridicule and retribution. And some of quantum physics' brightest minds concluded the phenomena are real.

Many famous quantum physicists showed interest in the paranormal.

I wasn't really aware of this fact when I wrote my first book, *Best Evidence*. But researching this new book, I was delighted to discover multiple quantum physicists also saw merit in investigating these phenomena. MIT professor David Kaiser devotes a full chapter in his book *How the Hippies Saved Physics: Science, Counterculture, and the Quantum Revival* to the topic of the implications of quantum physics for parapsychology. In it, he flags multiple famous quantum physicists who showed interest in the paranormal:

- Yale's eminent physicist Henry Margenau co-founded and edited the journal *Foundations of Physics* (its editorial board has included multiple Nobel Prize winners). Margenau urged the scientific study of extrasensory perception (ESP) and delivered a keynote address to the American Society for Psychical Research. "Margenau highlighted quantum entanglement as a likely means of reconciling ESP with action at a distance (ironically, Einstein himself used the analogy of ESP in discussing entanglement). As Margenau and Lawrence LeShan wrote, 'ESP is not stranger than some of the discussions that had recently emerged in quantum theory, such as Bell's theorem. Marching through one example after another, they found no contradiction between well-tested scientific laws and ESP.'"

- Nobel laureate quantum physicist Eugene Wigner is famous for introducing the hypothesis that human consciousness forms reality by collapsing the quantum wave function, reducing a world of probabilities to one outcome. "Wigner commented favorably in public and in print on CIA-funded research in the 1970s involving 'remote viewing' (clairvoyance) being conducted at Stanford Research Institute. Wigner concurred with [physicist Elizabeth] Rauscher that Bell's theorem seemed relevant, and encouraged her work on multidimensional space-times and psi phenomena."

- The root of today's quantum technology revolution is John Stewart Bell's 1964 theorem showing that quantum mechanics permits the seemingly impossible phenomenon of instantaneous connections between far-apart locations. In proving our universe is "non-local," his work radically upended a bedrock assumption of classical Newtonian 19th-century physics that said an object is influenced directly only by its immediate surroundings. Bell was sympathetic to parapsychology research. "When pressed to give an opinion on possible connections between parapsychology and his own work on quantum entanglement, John Bell refused to dismiss the matter out of hand.... Bell wrote that his experiences as a young physics student tempered his judgement of psi research. Critics of parapsychology often complained about the small number of times that researchers had successfully repeated a claimed psi effect. But perhaps that was not too different from Bell's own frustrations in his student laboratory in Northern Ireland, he wrote, where he had failed miserably to reproduce the well-known laws of electric attraction and repulsion ... because of the damp. Perhaps a similar confounding factor masked exotic parapsychological phenomena as well, making them difficult to reproduce at will. In any case, Bell closed, good scientists should certainly keep an open mind; physicists had been surprised by seemingly impossible phenomena several times before."

- Nobel Prize–winning physicist Brian Josephson invented the "Josephson junction" circuit, which today is found hardwired into everything from quantum computer prototypes to instruments that measure neural activity inside the human brain. He is a dedicated explorer of paranormal phenomena. "When the *New York Review of Books* ran a feature article in 1979 that was critical of efforts to use quantum theory to explain psi phenomena, Josephson teamed up with [physicist] Costa de Beauregard, [physicist Richard] Mattuck and [physicist Evan] Walker to write a feisty reply. . . . Josephson's passion for the topic has not wavered to this day. He directs a 'mind-matter unification' project at Cambridge University and vigorously defends parapsychology from naysayers."

Author Kaiser doesn't mention Wolfgang Pauli, but he's a Nobel Prize–winning quantum physicist who explored parapsychology and accepted the reality of psychokinesis. Pauli reported experiencing it frequently and personally, according to an article entitled "Quantum Mechanics and the Consciousness Connection," published July 16, 2012, by the American Association for the Advancement of Science (AAAS):

"Wolfgang Pauli (1900–1958), an Austrian-Swiss physicist, won the Nobel Prize in physics in 1945 for the 'Pauli Principle' involving spin theory. However, Pauli was also famous for what became known as the 'Pauli effect,' which manifested in the spontaneous breakdown of laboratory equipment whenever he entered the room. It also manifested in other ways, including mischievous events like chairs collapsing and train cars decoupling, to the detriment of those around him, but not to himself. At one point in his life he collaborated with psychiatrist Carl Jung to study synchronicity, in part because of the 'coincidences' surrounding his effect on matter. Not surprisingly, he believed in psychokinesis – the influencing of matter by thought – and that parapsychology was worthy of study. In fact, he was one of the first to consider and explore 'quantum metaphysics.'"

Parapsychology research has direct implications for the "hard problem" debate. In a lengthy article in the journal *Frontiers in Psychology*, published September 7, 2022, entitled "What if Consciousness Is Not an Emergent Property of the Brain?" psychology professor Dean Radin and colleagues cite over 130 scientific studies done on the phenomena of clairvoyance, telepathy, precognition, and xenoglossy. The studies "demonstrate non-local aspects of consciousness, perceiving information in a way that is not limited by our conventional understanding of time and space and that is not dependent on the brain function."

Maybe it's time for science to exhibit a touch of humility, and retire the term "paranormal" until it figures out what's "normal."

As biologist J.B.S. Haldane famously declared, "The universe is not only queerer than we suppose, but queerer than we *can* suppose."

Quantum physics is truly weird. Dare to dig deeper?

At times while researching this book, I felt like Alice in Lewis Carroll's surreal *Alice in Wonderland* when, at one point in the story, the comfortable, fixed rules of reality suddenly disappear – her body telescoping upward and her feet, far below, nearly shrinking out of sight. Alice exclaims, "It would be so nice if something made sense for a change." Amen to that. Quantum physics is severely intellectually disorienting. I'm guessing most readers only want to wrestle with the minimum amount of physics necessary to convince them an afterlife is possible.

But for my more adventurous readers, I've assembled below some additional books and articles dealing with quantum physics that I found particularly interesting during my investigation.

Ready to go down the rabbit hole? Before you jump, I suggest you swallow twenty-one pages from Dr. Dean Radin's book *Entangled Minds*.

In one short chapter, entitled "A New Reality," Radin provides a capsule history of quantum physics. He also explains key concepts in quantum physics – including wave-particle duality, the double-slit experiment, superposition, the Uncertainty Principle, Bell's Theorem, entanglement, the measurement problem, and the major "interpretations" of quantum physics.

I found particularly helpful his comparison of 19th-century Newtonian physics to 21st-century quantum physics, and how the two radically differ.

Like the Cheshire Cat's smile, our old 19th-century Newtonian worldview disappears before our eyes.

"Reality" – Newtonian physics assumes that reality at the most fundamental level is made of matter (hard, discrete, tiny particles with a precisely measurable size, weight, and defined position in space), and secondly, that this physical world of "things" is objectively real, "out there," existing independently of us humans. Quantum physics suggests that reality at the most fundamental level is not made of matter – reality exists as a state of potentialities and possibilities. And the fundamental properties of the world are not determined until they are observed by us. We

essentially bring the physical world of things into existence. Consciousness is fundamental, not matter.

"Locality" – Newtonian physics assumed that the only way objects can be influenced is through direct contact. But the quantum physics concept of "Non-Locality" – embodied in the quantum phenomenon of "Entanglement" – means that our commonsense assumption that ordinary objects are entirely and absolutely separate is incorrect. In unobserved states, quantum objects are connected instantaneously through space and time, even when they are not in direct physical contact.

"Causality" – Newtonian physics assumes that the arrow of time points only in one direction. Thus, cause always precedes effect, the present always precedes the future, and the past is gone forever. In contrast, quantum physics suggests that time is not fundamental, effects can precede causes, and the future can affect the past ("Retrocausality").

"Determinism" – As noted above, Newtonian physics assumes that only matter is real – little hard, discrete, particles that interact between each other in a fixed, predictable way; so science can predict the future completely if we know all the starting conditions and causal linkages. The future of everything – and everybody – in the universe is fixed and determined.

We're just blindly programmed meat robots. Newtonian physics is deterministic. Quantum physics challenges multiple Newtonian science assumptions underpinning its belief in Determinism, including the nature and reality of matter itself, the way causality works, and the ability of science to calculate the starting conditions needed to determine a future outcome with certainty. Quantum physics is all about probabilities.

As one cat disappears, another appears.

If Alice's grinning Cheshire Cat is curious, Schrödinger's feline friend is even "curiouser," suspended in a quantum state called "superposition." Colin Stuart explains the concept in his article "10

Mind-Boggling Things You Should Know About Quantum Physics," published in *All About Space* magazine, March 15, 2022:

"Objects can be in two places at once.... An electron, for example, is both 'here' and 'there' simultaneously. It's only once we do an experiment to find out where it is that it settles down into one or the other. This makes quantum physics all about probabilities. We can only say which state an object is most likely to be in once we look. These odds are encapsulated into a mathematical entity called the wave function. Making an observation is said to 'collapse' the wave function, destroying the superposition and forcing the object into just one of its many possible states."

The most famous illustration of this weird quantum physics concept is "Schrödinger's Cat," which is simultaneously alive and dead *at the same time*, until it is observed.

An article in *New Scientist*, entitled "Schrödinger's Cat," explains its history: "Devised in 1935 by the Austrian physicist Erwin Schrödinger, this thought experiment was designed to shine a spotlight on the difficulty with interpreting quantum theory. Quantum theory is very strange. It says that an object like a particle or an atom that adheres to quantum rules doesn't have a reality that can be pinned down until it is measured. Until then, its properties, such as momentum, are encoded in a mathematical object known as a wave function that essentially says: if you make a measurement, here are a range of possible outcomes. The inevitable question that arose as the theory developed was: what, then, is the thing doing before that? The most prominent answer in the 1930s came from the Copenhagen Interpretation, developed in the Danish city by luminaries of quantum theory, Niels Bohr and Werner Heisenberg. This says that there really is no definitive reality before the measurement, and the object is in an undefined state known as a superposition.

"Schrödinger's thought experiment probed how this plays out when a quantum object is coupled to something more familiar. He imagined a box containing a radioactive atom, a vial of poison and a cat. Governed by quantum rules, the radioactive atom can either decay or not at any given moment. There's no telling when the moment will come, but when it does decay, it breaks the vial, releases the poison and kills the cat.

"If the Copenhagen interpretation is correct, then before any measurement has occurred, the atom, and so also the cat, are in a superposition of being decayed/dead and not decayed/alive.... These days the thought experiment has taken on a kind of cult status.

There are Schrödinger's cat T-shirts, memes and hundreds of articles on the subject."

Quantum physics is an Alice in Wonderland world of the apparently illogical and nonsensical. Under the rules of this science, effect can precede cause.

A *New Scientist* article by Richard Webb entitled "Causality" explains our traditional, commonsense understanding of causality: "Things influence other things. That's a basic statement of any dynamic world where things change. . . . Causality is the study of how things influence one another, how causes lead to effects. In the classical world we live in, it comes with a few basic assumptions.

"The first big rule of classical causality is that things have causes. They don't just happen of their own accord. If a ball moves, the likelihood is someone kicked it; if an apple falls from a tree, it's because its weight became too great for the branch it was hanging from. Second, effects follow causes in a predictable, linear manner. You swing your leg, make contact with the ball, and off it moves, in that order and no other."

But is this always true? Not at the quantum level.

The article continues: "Unfortunately, matters of cause and effect get distinctly murkier in the other realm of modern physics, the quantum world. Unlike relativity, which governs the universe on large scales – the scales of stars, planets and galaxies – quantum theory determines the workings of very small things such as individual particles.

"For a start, the concept of uncertainty is hard baked into the quantum world. By setting certain limits on how accurately we can measure certain quantities at a quantum level, it also mucks with a conventional view of cause and effect. For example, one consequence is that pairs of particles can pop up at random from an empty vacuum, as long as they disappear again quickly enough not to violate the quantum uncertainty principle. As long as they exist, these 'virtual' particles can have effects on the real world – an apparent case of an effect without a cause.

"The quantum world is plagued by what Einstein called 'spooky action at a distance' – influences travelling seemingly instantaneously between particles, where measuring the state of one seems to influence the state of the other far faster than the speed of light. Experiments have even shown that two events can indeed seem to happen both before and after one another, opening the way on the quantum scale to what's called 'retrocausality' – the future influencing the present which can influence the past."

Author Kelly Oakes explores these shocking findings in her article published in the January 15, 2020, issue of *New Scientist* entitled, "In the Quantum Realm, Cause Doesn't Necessarily Come Before Effect." Her conclusion? "Playing fast and loose with causality does more than make for confusing mornings. It could shake physics to its very foundations."

It's already started. Spurred on by quantum experiments that scramble the ordering of causes and their effects, some physicists are figuring out how to abandon causality altogether.

Take the March 2021 *Wired* magazine article entitled "Quantum Mischief Rewrites the Laws of Cause and Effect." The author launches her story with a fun example: "Alice and Bob, the stars of so many thought experiments, are cooking dinner when mishaps ensue. Alice accidentally drops a plate; the sound startles Bob, who burns himself on the stove and cries out. In another version of events, Bob burns himself and cries out, causing Alice to drop a plate. Over the last decade, quantum physicists have been exploring the implications of a strange realization: In principle, both versions of the story can happen at once. That is, events can occur in an indefinite causal order, where both 'A causes B' and 'B causes A' are simultaneously true. 'It sounds outrageous,' admitted Časlav Brukner, a physicist at the University of Vienna.

"The possibility follows from the quantum phenomenon known as superposition, where particles maintain all possible realities simultaneously until the moment they're measured. In labs in Austria, China, Australia, and elsewhere, physicists observe indefinite causal order by putting a particle of light (called a photon) in a superposition of two states. They then subject one branch of the superposition to process A followed by process B, and subject the other branch to B followed by A. In this procedure, known as the quantum switch, A's outcome influences what happens in B, and vice versa; the photon experiences both causal orders simultaneously."

In Alice's wonderland, the white rabbit is always running by, clutching his watch, and exclaiming, "I'm late, I'm late, for a very important date! No time to say 'Hello, Goodbye' I'm late, I'm late!" In quantum physics' wonderland, he would never be late, because time doesn't exist.

"Many theoretical physicists have come to believe that time fundamentally does not even exist." So announced *Scientific American* magazine in its June 1, 2010, article entitled "Is Time an Illusion?"

Author Craig Callander masterfully explains the history of our idea of time: "Newton proposed that the world came equipped with a master clock. That clock idea gave the universe a sense of order, continuity, duration, a directional flow and order of events (past, present, future). The idea of a timeless reality is initially so startling that it's hard to see how it could be coherent. Everything we do, we do in time. The world is a series of events strung together by time."

SciTechDaily magazine, in its May 22, 2023, edition, notes that Albert Einstein was one of the first to challenge our everyday, commonsense view of time: "Scientists long assumed that time is absolute and universal – the same for everyone, everywhere, and existing independently of us.... But Albert Einstein's theory of relativity showed that time is relative rather than absolute – it can speed up or slow down depending on how fast you are traveling, for example."

As Einstein famously wrote to a scientist friend, "For us believing physicists, the distinction between past, present and future is only a stubbornly persistent illusion."

More recently, theoretical physicist Carlo Rovelli has developed a quantum physics theory that goes even beyond Einstein's relativistic "space-time" concept to propose a physics with no need for time at all. Andrew Jaffe, a cosmologist and head of astrophysics at Imperial College London, sums up Rovelli's theory in his *Nature* article titled "The Illusion of Time."

Jaffe writes: "According to theoretical physicist Carlo Rovelli, time is an illusion: our naive perception of its flow doesn't correspond to physical reality. Indeed, as Rovelli argues in his book *The Order of Time*, much more is illusory, including Isaac Newton's picture of a universally ticking clock. Even Albert Einstein's relativistic space-time – an elastic manifold that contorts so that local times differ depending on one's relative speed or proximity to a mass – is just an effective simplification.

"So what does Rovelli think is really going on? He posits that reality is just a complex network of events onto which we project sequences of past, present, and future. The whole Universe obeys the laws of quantum mechanics and thermodynamics, out of which time emerges. Rovelli is one of the creators and champions of loop quantum gravity theory, one of several ongoing attempts to marry quantum mechanics with general relativity. . . .

"Alongside and inspired by his work in quantum gravity, Rovelli puts forward the idea of 'physics without time.' This stems from the fact that some equations of quantum gravity . . . can be written without any reference to time at all."

Deep meditators sometimes report experiencing a subjective state of timeless reality, an eternal Now. With science increasingly taking the idea of a timeless reality seriously, these subjective experiences may eventually turn out to have objective validity as well. Noted Zen philosopher, author, and meditation explorer Alan Watts explains time in terms that echo Rovelli's interpretation of quantum physics: "Future is a concept – it doesn't exist. There is no such thing as tomorrow. There never will be, because time is always Now. That's one of the things we discover when we stop talking to ourselves [mental chatter] and stop thinking. We find there is only present, only an eternal Now."

Lewis Carroll had a fantastic imagination. Alice enters Wonderland by stepping through a "looking glass" (mirror) and discovering a completely new world on the other side. Albert Einstein stepped through the mirror of his own fertile imagination and discovered the Theory of Relativity, demonstrating that 19th-century Newtonian physics' belief that space and time are absolute and independent entities was wrong. But his imagination balked at the quantum physics' concept of "entanglement," calling it "spooky action at a distance." He remained on the Newtonian side of the mirror. Fortunately, other imaginative minds stepped through the looking glass, discovering that space (distance) is fundamentally an illusion.

Entanglement violates the classical, Newtonian-science principle of "locality," which states that physical systems cannot be affected by anything that is not in their immediate spatial vicinity. If you live in New York, you can't physically turn a doorknob in Honolulu. You're 5,000 miles away. Distance matters.

But not at the most fundamental level of reality – the quantum level. Author Andreas Müller explains the mind-boggling concept of entanglement in his article in *Phys.Org* entitled "What Is Quantum Entanglement? A Physicist Explains the Science of Einstein's 'Spooky Action at a Distance'":

"In the simplest terms, quantum entanglement means that aspects of one particle of an entangled pair depend on aspects of the other particle, no matter how far apart they are or what lies between them. These particles could be, for example, electrons or photons, and an aspect could be the state it is in, such as whether it is 'spinning' in one direction or another.

"The strange part of quantum entanglement is that when you measure something about one particle in an entangled pair, you immediately know something about the other particle, even if they are millions of light years apart. This odd connection between the two particles is instantaneous, seemingly breaking a fundamental law of the universe."

The weirdness of entanglement doesn't stop with distances in space. How about distances in time? Author Elise Crull reports that quantum entanglement works for time as well as space in her February 2, 2018, article published in *Aeon* entitled "You Thought Quantum Mechanics Was Weird: Check Out Entangled Time." As she writes: "Up to today, most experiments have tested entanglement over spatial gaps. The assumption is that the 'nonlocal' part of quantum nonlocality refers to the entanglement of properties across space. But what if entanglement also occurs across time? Is there such a thing as temporal nonlocality? The answer, as it turns out, is yes. Just when you thought quantum mechanics couldn't get any weirder, a team of physicists at the Hebrew

University of Jerusalem reported in 2013 that they had successfully entangled photons that never co-existed."

At the end of Alice's surreal adventures, she steps back through the looking glass and the rules of reality return to normal. The weirdness remains behind the mirror.

Similarly, until very recently, scientists believed quantum effects were limited to the world of the infinitesimally tiny. In standard physics textbooks, quantum mechanics is described as the theory of the microscopic world. It describes particles, atoms, and molecules, but ordinary Newtonian classical physics still rules on the macroscopic scales of apples, humans, and stars. You'll still find this written in physics textbooks and in popular science articles and voiced by science experts and commentators on TV.

But it's already outdated science.

In an article In *New Scientist* on April 12, 2023, entitled "A Macroscopic Amount of Matter Has Been Put in a Quantum Superposition," scientists announced that researchers have put a sapphire crystal containing quadrillions of atoms into a superposition of quantum states, bringing quantum effects into the macroscopic world. The crystal only had the mass of half an eyelash, so we're not yet talking about everyday objects. But researchers are already working on scaling up the size.

For better or worse, there's no stepping back through the mirror to our old, familiar, 19th-century scientific worldview.

I'd argue it's for the better.

Quantum physics arrives just as humanity – driven by a Newtonian science vision of ourselves – pushes itself and the planet to the brink of extinction.

We need a major mind shift to survive.

Nick, Gene, Me – and You

Quantum physics could be the mental medicine we humans need to make it to the 22nd-century without destroying ourselves. Its implications for humanity are optimistic, uplifting, and a powerful antidote to corrosive, modern nihilism.

If we do make it that far, I've got some nominations for "thank you" awards.

While researching this book, I had the great honor and pleasure of corresponding with physicist Nick Herbert.

Physicist Nick Herbert

Nick, now 87, is a colleague of physicist Fred Alan Wolf (who kindly wrote a cover blurb for my book *Best Evidence*) and a founding member of the celebrated Fundamental Fysiks Group, which met at the Lawrence Berkeley National Laboratory in California in the 1970s. Their freewheeling, maverick thinking forced mainstream physicists to pay attention to the strange but exciting implications of quantum theory for our understanding of reality. MIT professor and author David Kaiser recounts their exploits in his book *How the Hippies Saved Physics: Science, Counterculture, and the Quantum Revival.*

Nick's group penned textbooks used in physics curriculums, flagging the Alice in Wonderland weirdness as well as the optimistic (for humans) discoveries embedded in the dull math. Notes Kaiser, "Today's undergraduates at MIT learn about Bell's Theorem in the first semester of quantum mechanics. That simply wasn't true for a long time. Questions about what it all means now have a place in the curriculum."

Members of the group also wrote popular books on quantum physics that were assimilated into the public psyche. Physicist Fritjof Capra's 1975 bestseller, *The Tao of Physics: An Exploration of the Parallels Between Modern Physics and Eastern Mysticism,* dared to explore the implicit links between quantum physics, philosophy, and spirituality that 19th-century Newtonian science refused to examine. It went on to sell over 1 million copies and be translated into twenty languages.

Nick himself wrote three physics books, but also made a name as a poet. He played a prominent part in the early Boulder Creek (California)BistroScene poetry movement, which resulted in two published chapbooks: *Physics on All Fours* and *Harlot Nature* (Nick also experimented with LSD and lived life to the fullest. See writer John Horgan's delightful, must-read interview with Herbert in the October 8, 2018, issue of *Scientific American*, entitled "Chasing the Quantum Tantra").

One of Nick's poems includes this humorous, clever, quantum-inspired stanza playing with the concept of wave-particle duality.

> *This is the World of the Quantum Mechanic*
> *Not the Butcher nor Baker nor Cook*
> *It's possibility waves when unregarded*
> *It's actual particles whenever you look*
> *In utter darkness safe from leerers*
> *Huge Waves of Maybe surged and swam*
> *But when I turned to look at them*
> *They turned into little bits of AM.*

Back in 1987, Nick Herbert masterfully explored eight different quantum physics interpretations in his classic book, *Quantum Reality*. His popular science book explained quantum physics without relying on advanced mathematical concepts. It sold over 100,000 copies, not including German and Japanese translations, and was required reading for physics undergraduate courses well into the mid-1990s (it's still available on Amazon).

In the closing chapter of *Quantum Reality*, Nick declared, "Science's biggest mystery is the nature of consciousness." Four decades later, Nick is still searching for answers. He sees flaws in every interpretation currently offered. As he explained in his email to me, "We are still really in Deep Mystery when it comes to quantum reality."

Nick shared with me an amusing story about his encounter with Nobel laureate quantum physicist Eugene Wigner: "Shortly after my book was published, I attended a big conference on quantum questions in New York at which Eugene Wigner spoke and his 90th birthday celebrated. After his talk, I went up and presented him with a copy of my book. He took one look and, in his Hungarian accent, he said: 'Quantum Reality? But is that not a contradiction in terms?'"

It was a witty reply that both men could fully appreciate.

Wigner and Herbert both spent their careers exploring a quantum physics funhouse where the laws of nature clearly appear to contradict reality at every step. Both men fearlessly followed that surreal path to its triumphant end – a revolutionary new science that humanity is just beginning to intellectually understand and psychologically embrace.

Researching this book, I came away with tremendous respect for Wigner and his fellow founders of quantum physics for the sheer audacity of their thinking, and their intellectual honesty in accepting the bizarre but logical conclusions demanded by the evidence.

I feel equal admiration for Nick and his colleagues at the Fundamental Fysiks Group who finally forced mainstream physics to stop ignoring the full implications of the mind-bending discoveries made by Wigner's generation.

Quantum physics brilliantly brings together matter, mind, and spirit into one unified, modern, 21st-century scientific worldview.

It does so without resorting to religious dogma – or to the equally dogmatic 19th-century assumption of scientific materialism. Beyond the discovery that consciousness is fundamental, and an afterlife is scientifically logical and possible, it leaves us free to choose where to go from there – to follow whatever spiritual or philosophical path each of us personally finds most compelling or true.

We humans created science. It's among humanity's greatest achievements. But science without spirit is sterile, incomplete, and harmful to both people and the planet. Nineteenth-century Newtonian scientific materialism gave us some amazing technologies and toys, but removed meaning and purpose from life and the universe. It was bad science and a bad bargain.

Quantum physics offers us a new scientific worldview infinitely more beneficial for both humanity and the Earth.

One Last Request….

If you've enjoyed this book, please consider posting a review on Amazon.Com
Thank you from the author!

ABOUT THE AUTHOR

Investigative journalist Michael Schmicker started his career as a crime reporter for a Dow-Jones suburban newspaper in Connecticut. He worked as a freelance correspondent in Southeast Asia during the Vietnam War and as an Op-Ed contributor to *The Asian Wall Street Journal*.

A nationally known writer on frontier science, Michael is the co-author of *The Gift, ESP: The Extraordinary Experiences of Ordinary People* (St. Martin's Press (USA)/Penguin Random House (UK). His first book, *Best Evidence*, has emerged as a classic in the field of scientific anomalies reporting. Michael has reviewed books for the *Journal of Scientific Exploration*, contributed to *EdgeScience* magazine, and served on the Board of Advisers for the Rhine Research Center.

His interest in investigating the paranormal began as a Peace Corps Volunteer in Thailand, where he first encountered a non-Western culture and people who readily accept the reality of ghosts and spirits, reincarnation, and other persistently reported phenomena unexplainable by Newtonian science. He spent his first year in Bangkok teaching English to middle school students at the royal Buddhist monastery of Wat Bowonniwet, and the subsequent two years writing and producing with Thai colleagues a *Sesame Street*–inspired educational television series for the Thai Ministry of Education. Before joining Peace Corps, he studied documentary film production at New York University and the British Film Institute. He earned a B.A. in philosophy, and a master's in educational communications (television) from the University of Hawaii.

Michael is author of the Amazon Top 100 historical fiction best-seller *The Witch of Napoli,* based on the true-life story of Italian medium Eusapia Palladino. On March 6, 2015, it made the Amazon Top 100, ranking #41 in paid books out of 3.3 million books available in the Kindle Bookstore. That same day, it hit #1 in both the Historical Fantasy and Victorian Historical Romance categories in three countries simultaneously – the US, Canada, and England. It spent 68 consecutive weeks as a Top 20 Best Seller in the Italian Historical Fiction category. Amazon subsequently selected *The Witch of Napoli* as one of 25 Historical Romance Bestsellers to be featured by Amazon's new Prime Reading program.

Visit him at MichaelSchmicker.com.

ACKNOWLEDGEMENTS

The author would like to especially acknowledge the help of his family, including son Christopher; his brother John, and John's wife Annette; and his sister Jody, for their reading of this manuscript, and their valuable insightful comments, questions, and feedback. A thank you also to Eleanor Duncan of MRM/McCann for her proofreading of this manuscript.

PHOTO/ART CREDITS

Book cover design: ACD/Andy Carpenter Design

Cover drawing and interior line drawings: Asseltaii/Fiverr

Flammarion engraving: Public domain, PublicDomainReview.org

Werner Heisenberg headshot: Wikimedia Commons

Niels Bohr headshot: Wikimedia Commons

Erwin Schrödinger headshot: Wikimedia Commons

Brian Josephson headshot: Wikimedia Commons

Max Planck headshot: Wikimedia Commons

Eugene Wigner headshot: Wikimedia Commons

Theater: Bing AI

Sherlock Holmes: Bing AI

Raymond Moody, MD: Amazon

Out of body art: Bing AI

Peter Fenwick, MD: YouTube, "q&a peter fenwick on death, consciousness and the afterlife"

Pim van Lommel, MD: YouTube, "The Mystery of Perception During Near Death Experiences - Pim van Lommel"

Jeffrey Long: Jeffrey Long

Blind baby: Bing AI

Sam Parnia, MD: New York University Media Relations

Woman's head with consciousness: Bing AI

Woman smelling flower: Peter Kratochvil/public domain

Federico Faggin: YouTube

Woman kissing baby: iStock Photo

Two men and wine bottle: Bing AI

Radio: Google Images

Falling dominoes: CanStockPhotos.com

Disappearing atom: Bing AI

19th-century scientific tools: Google Images

19th-century study of human vs. monkey brain: Google Images

Marcelo Gleiser: Marcelo Gleiser

Neuron: Wikimedia Commons

Marching scientists: Bing AI

Starry sky and watchers: Kendall Hoopes/Pexels

Freeman Dyson: Wikimedia Commons

John Wheeler: Sandia National Labs

Scientist with hands over eyes: Bing AI

Winged soul rising from the grave: Bing AI

ESP blackboard: Bing AI

Alice in Wonderland vintage book cover: Google Images

Cheshire Cat: Google Images

Schrödinger's cat: Bing AI

Soccer player: Bing AI

Melting clock: Bing AI

Rocket ship in space: Bing AI

Nick Herbert headshot: Reno DeCaro

Printed in Great Britain
by Amazon